ONE GIANT LEAP

Apollo 11 Forty Years On

Piers Bizony

Essex County Council Libraries

First published in Great Britain
2009 by Aurum Press Ltd
7 Greenland Street, London NW1 0ND
www.aurumpress.co.uk

A catalogue record for this book is available from the British Library.

ISBN 978 1 84513 422 8

1 3 5 7 9 10 8 6 4 2
2009 2011 2012 2010

Book design by Piers Bizony

Originated and printed in Singapore

CONTENTS

FOREWORD

WHAT IMAGE COMES to mind when we think of Apollo 11? We all know that famous picture of "the first man on the moon," don't we? There's a golden visor staring out at us, like the cold and passionless eye of some Cyclopean robot. The visor inside its bulky helmet caps a stiff-looking spacesuit. Fat white hoses snake in and out of blue and red valve connectors on the spaceman's chest, and the balloon-like arms of this clumsy creature are tipped by gray gloves with fingers like stubby sausages. The background is a flat expanse of gray, powdery soil and a pure black sky. There are no sepia tones to age the shot, no long-vanished clothing styles to help us date roughly when the picture might have been taken. The person in the spacesuit looks like any other person in a spacesuit, because you hardly ever see their faces inside those sunproof golden masks, and spacesuits still look pretty much the same today as they always have. The crisp, clean, high-quality color photo looks as if it might have been taken last year, last month, yesterday. The truth—that it was taken forty years ago—seems impossible. Stranger yet is the fact that no good photographs exist of the first man on the moon actually walking on it. What we see, in that eternally famous and almost perfectly composed picture of the first man on the moon, is the second man.

All the lunar astronauts deserve heroic status, and each Apollo mission was just as important as any other. All the flights were dangerous, ambitious, extraordinary. But we have a relentless fascination for people who do things for the very first time. The names of Neil Armstrong and Buzz Aldrin are memorable to us because they were the first men to land on the moon, and many people also recall Michael Collins, orbiting overhead in the command module, but only those space enthusiasts with a pedant's grasp of detail can recall exactly who commanded Apollo 14, or which astronaut stayed aboard Apollo 16's mothership while the other two went down to the lunar surface. History can be cruel, yet we cannot blame ourselves for concentrating on the highlights of events that, in many cases, happened while we were young, or even before we were born. Our minds are overloaded with what's happening today, never mind the details of events long past. This book is a celebration of Apollo 11, but it has to be considered as standing for the entire Apollo adventure, not just that one mission. That said, there is something unique about the photos taken during Apollo 11. Most of the astronauts who came afterwards had the luxury of spending more time on the lunar surface, and there were more opportunities for them to pause for a moment and take photos of each other. The bulk of the many fabulous images gathered together in countless Apollo retrospective books and exhibitions come from those later missions.

For the crewmen of that very first landing, taking photographs took second place to simply completing their mission without making mistakes that might kill them in their still relatively untested spacecraft. Even so, Armstrong and Aldrin had cameras strapped to their chests throughout their time on the surface: Hasselblads loaded with 70mm roll film. Orbiting overhead in the command module, Michael Collins was equipped with similar cameras. Nearly 600 frames were exposed during the mission. Yet in most books recounting this greatest of all space adventures, only a handful of the images tend to appear. Why is this?

Partly, it has to do with the fact that the Apollo 11 crew weren't always at the top of their game when it came to making beautiful photographic works of art. Many shots, and especially those taken inside the spacecraft, are blurry, under-exposed, or so random in the choice of subject matter that they can only have resulted from the shutter button being pressed by accident. For four decades, picture editors have looked at these apparently disappointing failures and set them aside. Now, looking at them with a fresh eye, we gain a vivid impression of men busy with urgent tasks; men unable, or unwilling, to concentrate for more than a few brief seconds at a time on taking photos; hurtling towards the moon at 25,000 miles per hour while checking that none of the ten dozen switch settings on their control panel hides the one incorrect flip that might kill them; men pressing strings of numbers into their computer, knowing that one wrong digit out of a hundred could spin them to their deaths in the wastes of space. The crew of Apollo 11 were trying to avoid potentially fatal errors among a thousandfold tasks during their eight-day voyage. Is it any wonder that their photos, snapped in very rare spare moments, were sometimes less than perfect?

If the task of getting to the moon occupied almost every waking second of the astronauts' time, then exploring its surface was just as demanding. In fact one of the great failures of this otherwise magnificent project was NASA's misplaced zeal for "professionalism" during every moment of the moon walk. Senior managers were worried that taxpayers would complain if the astronauts took time out from collecting rocks or digging for soil samples. Consequently the moon walkers' schedules were planned out in advance almost literally to the minute. There were various boxes of equipment to unwrap from the sides of the lunar module and a long list of scientific experiments to deploy in a certain order. Then there was "x" amount of terrain to cover and "y" pounds of rock samples to be gathered in a given time, and so on.

Those photos that were officially scheduled to be taken were of the moon itself: the texture of the soil, the locations of rocks and boulders, the geological (or, more properly, selenological) character of the immediate surroundings and far horizon. Of the two Hasselblad pictures that Aldrin took of Armstrong on the moon, one is almost comically incorrect, showing just a sliver of Armstrong's suit as he disappears out of the frame. Of course, some shots, and especially those of Aldrin taken by Armstrong, do work stunningly well, and it is these that have tended to dominate our cultural memories of Apollo 11 over the last four decades. Armstrong and Aldrin had to steal the odd moment or two from their schedules to look around and absorb the impact of what they were doing. No wonder they found it so hard to tell the rest of us about it afterwards—hard, perhaps, to tell themselves.

In some sense, the stars of these photos are the machines, not the men. Apollo really was a portent of the future: human flesh was concealed behind—or rather, inserted deep into—the convoluted tangle of Apollo's "man-machine interface." For historians and psychologists alike, the emphasis on shots of control panels, lunar lander legs, pieces of reflective foil, docking probes, and other pieces of equipment, and the relatively infrequent appearance of human faces, is telling. Man is here, in each and every picture, but represented by his fascination for the vehicles that he knows how to build and fly, rather than by Man himself.

The crew of Apollo 11 were professional, cordial, but not particularly intimate with each other, and sometimes it shows, although Buzz Aldrin's portrait of Neil Armstrong safely back in the lunar module after the moon walk is the very essence of generosity: an acknowledgement that the first man on the moon deserves to smile while the second sets his disappointment aside and takes his picture for posterity—perhaps aware, in retrospect, that he didn't really shoot enough of Armstrong while the two men were actually walking on the moon.

There was no special provision in Apollo 11's flight plan for taking good publicity photographs of "men on the moon," despite this mission supposedly being of such historic significance. The fact is that the global media machine of 1969 was not the instantaneous and all-consuming entity that it is today. Only at the last minute did NASA recognize the importance of those grainy television broadcasts of the moon walk—and the space agency almost entirely missed the importance, to people back on earth, of seeing photos of humans on the moon, not just rocks and soil, machines, and instruments.

Armstrong and Aldrin's familiarization with the TV cameras and Hasselblads was imperfect at best, a late and largely unwelcome addition to their training schedule. It was common for most astronauts to complain that photos and video footage got in the way of more urgent tasks. These men were test pilots trying to stay alive, not photographers or film directors trying to make art. Understanding all this, we can now look at a wide selection of Armstrong, Aldrin, and Collins's photographs with a fresh perspective, and see the hidden qualities of certain shots that might once have been regarded as "failures."

This book gathers together more images from Apollo 11 than any mass-market publication has attempted before. In gathering them, I called on many years of research, image cataloguing, and digital scanning conducted by Apollo historians, some of them working within NASA's History Office, but many more, it has to be said, dedicating private time and personal resources to help make the historical record of Apollo as complete as possible. I would like to thank: Paul Fjeld, Ed Hengeveld, Eric Jones, Mike Marcucci, Jack Pickering, and Kipp Teague, and from NASA, Steven Dick, Colin Fries, Mike Gentry, Connie Moore, and Margaret Persinger.

One view that we seldom see in the photos is of both Apollo craft docked together; and even when we do see the vehicles, the lighting conditions of deep space are such that the glare of reflected sunlight from one side of a ship dazzles the eye, while shadows obscure details on the other flank. I am grateful to an outstanding computer graphics artist, John Ortmann, for his digital artworks showing exactly what the spacecraft of Apollo 11 looked like, down to the last rivet and crinkled piece of foil. John Knoll, co-creator (with his brother Thomas) of Photoshop, has also made for this book a superb interior view of the lunar module's cabin. This is surely the closest that any of us will get to sharing the view out of Armstrong's window as he avoided a dangerous crater in the last few seconds before touchdown. David Mindell, Professor of the History of Engineering and Manufacturing at the Massachusetts Institute of Technology (MIT) has also been generous in giving me advice about Apollo's computer systems.

In the UK, I received valuable advice from Michelle Clarke at the Institution of Engineering and Technology (IET), and other support from her colleagues Dickon Ross and Vitali Vitaliev in the IET's publications division. Chris Riley, co-producer of the oustanding Apollo documentary film *In The Shadow Of The Moon*, was helpful with advice and encouragement. Thanks also to my old friend and constant advisor Doug Millard at the Science Museum in London, and to Michelle's husband Adrian, a true and generous Apollo fan. Finally—although I really mean first and foremost—I thank my wife, Fiona, and my family, for being the best "ground support" team that any space writer could have. Our two dogs and one cat, being incapable of reading, will simply have to accept my gratitude in the form of biscuits.

1 SETTING THE SCENE

SETTING THE SCENE

1

THE STORY OF HOW AND WHY we sent men to the moon has been told many times. This book is a series of themed and illustrated essays about Apollo's cultural legacy viewed from the perspective of today, rather than a detailed history of events. Even so, it is probably worth reminding ourselves of the events leading up to that adventure:

In July 1969 an incredible machine thundered towards the moon. Apollo 11's Saturn V rocket highlighted the best and the worst of our desires. It satisfied our yearning to explore space in peace, but its design was inspired by wars on earth. Half a century ago, two brilliant engineers triggered a rocket competition between Russia and America. Each man had known the worst cruelties of his times. One narrowly avoided death in a labor camp, and the other may have been partially responsible for creating one. Both men designed and built rocket weapons of devastating destructive power, even as they propelled us into the cosmos.

In 1938, aircraft engineer Sergei Korolev was developing simple rockets at an army laboratory in St Petersburg when he fell victim to one of Stalin's terror purges. He was savagely beaten and tortured, then imprisoned in a freezing Siberian gulag. Three years later, on the brink of death, he was ordered to Moscow. The German army had invaded and Stalin suddenly needed engineers. In 1945 Korolev was sent into Germany to recover the remnants of rockets similar to those that he had dreamed of building. He found that a young German pioneer had already outstripped his wildest ambitions.

Wernher von Braun is probably the only "rocket scientist" whose name is familiar to everyone. Born into an aristocratic family, he was obsessed by space. When the Nazis came to power he persuaded Hitler that rockets would make effective weapons. He and his dedicated team developed the V-2 "Vengeance" rocket at Peenemünde, on the Baltic coast. During the war, a nightmarish underground factory was built at Nordhausen, where half-starved slave laborers

assembled the rockets under SS supervision. In the last months of the war, British, American, and Russian intelligence teams scoured the devastated German heartland for any remains of the V-2. This was, after all, the world's first guided missile, and its descendants would define the future balance of power. Competition among the teams was ruthless, even though they were supposed to be allies. Von Braun decided the Americans were his best hope, because they would probably employ him. The British couldn't afford him and the Russians might shoot him. He staged a brilliant escape for himself and his closest colleagues under the noses of SS squads, who were by now indiscriminately killing "disloyal" Germans.

There followed a decade of half-hearted V-2 tests in the White Sands desert of New Mexico. Von Braun was frustrated by the US government's lack of interest in him (he was rather an embarrassment, as well as a potential asset), so he became a public campaigner. In the late 1950s Walt Disney promoted him as a trustworthy crusader for space exploration. Korolev, meanwhile, created Raketa-7 (R-7), the first intercontinental ballistic missile, along with a vast and secret launch complex in the remotest reaches of Khazakstan. The R-7 was a weapon. Its purpose was to hurl heavy nuclear warheads across the Atlantic and onto America, but Korolev knew it could also carry lighter payloads all the way into space. On October 4 1957 he proved it by sending a satellite called Sputnik into orbit. A larger payload, Sputnik II, flew a

Almost like some sacrificial victim, Gordon Cooper is assisted into his capsule prior to flying a new and relatively untested rocket.

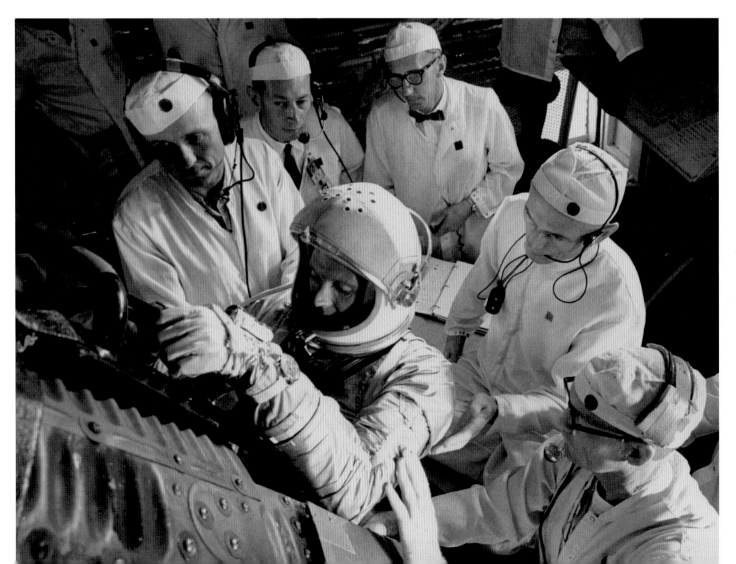

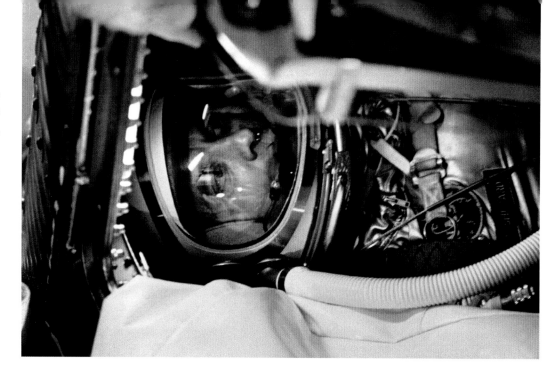

America's first man in space, Alan Shepard, prepares for launch in his tiny Mercury capsule.

similar trajectory on November 3, carrying a passenger, a dog called Laika. Korolev's intentions were clear enough. He wanted to launch a man. He now told the Red Army generals that he could build a satellite to spy on the West, but first he would have to enlist a pilot with excellent eyesight to look out of the satellite's window and check on what the spy cameras might see. The generals believed him, and in October 1959 a squad of "cosmonauts" was formed.

In America, President Dwight D. Eisenhower was reluctant to spend money on rockets. He feared the influence of what he called "the military–industrial complex". Only when a small-scale Navy satellite launcher called Vanguard blew up on the pad was von Braun summoned, at last, to rescue America's faltering space effort. The National Aeronautics and Space Administration, more usually known by its acronym NASA, was created in July 1958 and given just enough cash to develop a small manned space capsule, the Mercury. Eisenhower fervently hoped that what he dismissed as "all this Buck Rogers nonsense" would end soon.

THE RACE BEGINS

Von Braun and his colleagues, under the leadership of NASA's first chief administrator, Keith Glennan, and his hard-driving successor James Webb, were hard-pressed to catch up with their Russian rivals. On the evening of April 12 1961 America was extremely disturbed by yet another triumph for the R-7 missile. Yuri Gagarin flew into space aboard a spherical capsule called Vostok. A newly-elected president, John F. Kennedy, was desperate to respond. "If somebody can just tell me how to catch up. Let's find somebody. I don't care if it's the janitor over there, if he knows how." Three days later, a CIA-backed raiding party landed at the Bay of Pigs in Cuba, hoping to destroy Fidel Castro's regime. The invasion was a disaster, and Kennedy turned to space in a bid to restore America's international prestige.

On April 20 Kennedy asked his vice-president, Lyndon Johnson, an ardent advocate of US progress in rocketry, "Do we have a chance of beating the Soviets by putting a laboratory in space, or by a trip around the moon, or by a rocket to land on the moon? Is there any other space program which promises dramatic results in which we could win?"

John Glenn became the first American to orbit the earth,
propelled into space by an Atlas missile (right).

A final decision hinged on NASA literally getting their manned space program off the ground. On May 5 Alan Shepard was launched in a Mercury capsule, atop a small Redstone missile designed by von Braun. Shepard's flight was just a fifteen-minute ballistic hop, a mere fraction of an orbit, but it was enough to boost Kennedy's confidence. In a historic speech on May 25 1961, he told Congress, "I believe that this nation should commit itself to achieving the goal, before this decade is out, of landing a man on the moon." NASA's Apollo project, one of the greatest engineering achievements in history, was kicked off because America's first man in space was launched just twenty-three days after Russia's first man. Shepard should have flown in March 1961, but there had been last-minute problems with the Mercury. If it hadn't been for that slight delay, an American would have been first into space, and Kennedy might never have felt the need to send men to the moon.

With Soviet politicians eager to respond to Kennedy's dramatic move, Korolev was forced to adapt his Vostok to carry two cosmonauts in the same small capsule, and even three if they gave up their spacesuits. A veil of secrecy prevented the outside world from learning details of the hardware. Over the next five years, Russia appeared to stroll from one victory to the next: the first woman in space, the first multi-man crews, the first space walk, and even the first stab at an orbital rendezvous between two ships. These were real achievements, but they stretched the Vostok design to its limits. Korolev was exhausted. On January 14 1966 he died during what should have been a routine stomach operation. It turned out he was riddled with cancer, aggravated by old injuries from his time in the gulag. He had been planning the Soyuz, Russia's workhorse capsule which is still in use today. He had also embarked on a giant lunar booster, the N-1. Now his legacy was in the hands of less capable administrators. The N-1 was a disaster, and Russia never came close to landing a man on the moon, although there was some concern, by 1968, that it could, perhaps, send a single cosmonaut around the moon using an intermediate-sized rocket, the Proton, thereby stealing Apollo's thunder. NASA could not be sure about any of this at the time, because the Soviets were so secretive.

The Apollo project continued at breakneck pace, incurring its own terrible risks. In January 1967, three astronauts, Gus Grissom, Ed White, and Roger Chaffee, died in a launchpad fire while testing systems aboard the first Apollo capsule. After a tense two years of recrimination and redesign, NASA recovered its momentum. Apollo 8 carried Frank Borman, James Lovell, and William Anders around the moon during Christmas 1968. Seven months later, in July 1969, John F. Kennedy's pledge to land a man on the moon "before this decade is out" was fulfilled when Apollo 11's lunar module Eagle landed Neil Armstrong and Buzz Aldrin safely on the surface, while Michael Collins waited in orbit aboard the command module Columbia.

In 1975 President Richard Nixon called a halt to the space race. He sent the final Apollo craft to rendezvous in orbit with a Soviet Soyuz capsule, and today the fruits of that union can be seen in the International Space Station. An adventure born out of fear and rivalry now helps to maintain peaceful industrial and scientific cooperation between Russia, America, Japan, Canada, and the European Union.

2 ROCKETS & CATHEDRALS

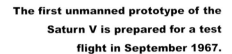
The first unmanned prototype of the
Saturn V is prepared for a test
flight in September 1967.

ROCKETS & CATHEDRALS

2

IT'S A CLICHÉ TO COMPARE the tall white spire of the Saturn V to a modern cathedral, but the thing about clichés is that they usually have some truth in them. The medieval castles were impressive statements of power and prestige, yet they were brutishly practical, and not nearly so much of a challenge to build—and, of course, not so joyously beautiful—as the extremely tall, very wide, and pointlessly ornate cathedrals, which were designed to show off what could be done with stone, wood, and glass when those materials were stretched to their limits. For all the practical reasons of Church and State that ensured their funding, the simple truth is that cathedrals rose up from the ground because communities wanted to express the utmost that they could achieve. But on a practical level, the communities who built them learned the art of managing large-scale construction projects, one of the foundations of modern civilization. Building cathedrals was—just as John F. Kennedy said of Apollo—"difficult and expensive to accomplish." Their scale and complexity required radically new techniques in architecture, mathematics, and structural engineering, and many other disciplines that were not apparently strictly needed for the everyday work of putting up houses and barns. No one can doubt the craft skills of medieval house-builders, nor the literally vital importance of the farmers' craft, but sometimes our ancestors wanted to do more than merely live.

Castles and forts exploited vast tonnages of stone and brick to deliver protective strength. They were rooted into the ground like vast boulders. In vivid contrast, cathedrals used as little stone as the masons felt they could get away with, in order to achieve an impression of lightness, taking to the air, ascending to the heavens. They may have been religious structures built by mortal men in honor of God, but they had a prideful, defiant character, too: a refusal to accept the apparent limits of Nature, and a desire to push a little further and see what the *real* limits might be. There was no stopping the push towards further extremes of height and scale, expressed with the strongest yet lightest materials. It was only a matter of time before some cathedral or other would stretch its spire indefinitely high into the sky.

The huge Vertical Assembly Building takes shape (left). Notice the three mobile launch towers. The technology of the gigantic Saturn V (above) was expected to have a long future.

The Saturn V launch vehicle for the Apollo project shared some of these qualities, especially in its expression of towering scale and strength derived from the lightest materials; and, just like the cathedrals, it was an expression of defiance against limitations. To take just one example: the people of Apollo took the most dangerous chemicals they could find and harnessed them, like sled drivers leashing a squad of snarling lions instead of tame huskies. Why? Partly because the unforgiving mathematics of the lunar flight demanded the use of lightweight hydrogen fuel, but also because they wanted to see if they could get away with it. The kerosene fuel used for the first stage was familiar to all aviation workers, but the super-cold liquid hydrogen in the upper stages was something new and terrible. It played havoc with the fabric of any structure that tried to contain it, and exploded with the power of a small atom bomb at the slightest spark. The tiniest variations in the thickness of welded seals in a metal tank were exaggerated into dangerous flaws by the thermal shock of liquid hydrogen pouring onto them. And this is the fuel that they chose for the Saturn: a substance that burns in the heart of the sun.

As if this wasn't enough of a challenge, hydrogen was only one of the dangerous chemicals that the Saturn had to contain. The rocket's combustion process also needed oxygen. To the inexact and non-technical imagination, liquid hydrogen stored at 300 degrees below zero seems pretty much "as cold as" liquid oxygen in an adjacent tank. In terms of the underlying physics, there's a 100-degree difference in temperature between them: as great as that which separates a pan of boiling water from a tray of ice cubes. Yet these fluids in their tanks had to share a common dome-shaped bulkhead, barely an inch thick, to make the most efficient use of all the space inside the stage. To cap it all, the structure as a whole had to be ridiculously, impossibly lightweight. According to NASA's brief, the metal skin, all the pipework, and the massive engines had to comprise no more than seven percent of an upper stage's entire

What a rocket should look like: a test version of Saturn V on the pad (right).

The monstrous first stage for a Saturn V takes shape.

weight when fuelled. At the same time, the thin shells of these metal cylinders had to withstand a thrust of six million pounds from below, while supporting the weight of yet another 130-ton third stage and lunar ship above.

Yes, this was a cathedral of sorts: a thunderous defiance against Nature, a prideful tower. But Saturn V was a prayer of humility too. It was something vastly greater than anything that one man or woman working alone could have created. So long as human civilization persists, the Saturn V will be remembered, long after the names of all the thousands of people who built it have been forgotten. The space visionary Arthur C. Clarke couldn't have put it better when, in 1972, he wrote, "An age may come when Apollo is the only thing by which people remember the United States, or even the world of their ancestors, the distant planet Earth."

THE HEAVENS AND THE EARTH

A kind of spiritual urge played a role for both sides in the Cold War space race. The genuine joy that Russians felt about putting Sputnik and Gagarin into orbit was tied to a sense of what those achievements promised: that the Communist dream might actually be possible, despite the dreadful hardships that this colossal social experiment had yielded so far. Soviet planners dreamed of taming Nature in the service of Man. Sputnik suggested that even the stars themselves might be tamed. If the Russian people's long-held religious faith had been denied by the earthly politics of Marx, Lenin, and Stalin, now it was free to express itself in a more materialist kind of language. What were rockets but new cathedrals? What was the Sputnik if not a polished metal prayer? The cosmos was an analog of heaven that

> "An age may come when Apollo is the only thing by which people remember the United States, or the world of their ancestors, the distant planet Earth."
>
> **Arthur C. Clarke, 1972**

ordinary Russians were allowed to dream of reaching. There was even a new kind of god to worship. The Kremlin insisted that the identity of Sergei Korolev, the architect of Russia's early space triumphs, should remain a secret for fear that American spies would assassinate him. Perhaps by accident, but surely to the Kremlin's satisfaction, this measure created an anonymous, faceless deity, "The Chief Designer," who inspired love and awe among Russians and an eerie sense of dread in America. Lethal tensions that had existed between the Soviet State and the Russian Orthodox Church briefly found some respite among the rockets.

Meanwhile, NASA hitched its wagon to America's favorite dream. On the new space frontier, astronauts were the latest bunch of cool-hand cowboys heading off in a thunder of hooves and a cloud of dust towards some new horizon. In the classic pioneering myth, a small fleet of ships sets sail across an uncharted ocean to discover new territory; a settlement is established at the point of landing; a settlement becomes a town. In time, other ships arrive with families of settlers to populate the new territories. Inspired by Wernher von Braun, NASA's champions talked of "colonizing the space frontier." Always, their plans were worked out with the most exact attention to fuel weights, rocket thrusts, orbital heights and speeds. As Arthur C. Clarke has pointed out, "No achievement in human affairs was ever so well documented before the fact as space travel."

An eerie machine for a surreal project: a lunar landing simulator suspended from a gantry on wire cables.

The flaw in the plan was simple. Where was the trade? Ships could be sent into orbit or to the moon, but what could they be expected to bring back as a return on the investment required to build them? What gold or silver could the space cowboys stash in their saddle bags to pay their way over the next horizon? This fundamental economic problem could not be wished away by any amount of romantic language. Even in the 1950s NASA's engineers had proved, on paper, that a flight to the moon was possible, and that there were many good scientific and spiritual reasons for attempting one; but there was no obvious way to justify spending the vast sums of money required to build the rockets.

Cathedral-building sent us moonwards. In the wake of Sputnik, the economic arguments against rocket adventures became strangely inverted, so that the drawbacks became, instead, the justifications. By accepting the huge expense of these abstract adventures without any prospect of a sensible financial return, both Russia and America set out to show what their rival societies might be capable of on earth: how much *more* they could do than merely running their domestic economies or equipping their armies. They built rocket–cathedrals because there was only so much pride that two global superpowers could take in the unpleasant business of war, or the dull, unremarkable process of simply keeping their people alive. They aspired to more—and so did a great many of their citizens.

It is essential for us to realize that neither of the Cold War superpowers would have reached for space—let alone the moon—if that prize hadn't meant something to the millions of ordinary people they were trying to impress. The moon sat at the top of the tallest spire that anyone could think of building. So they built it. And it didn't fall down. All that happened in later years was simply that people stopped going to that particular church.

THE HERETICS

Of course, not everyone was sold on the Apollo dream. In the late 1960s the counterculture was in full swing, and for the hippy generation NASA represented something devilish: The Man, that figurehead of the White Anglo-Saxon Protestant (WASP) capitalist élite, crushing human instinct in service of the profit motive, subverting Nature to his will and creating a global army of robot zombies. Norman Mailer, the famed writer and self-appointed spokesman for a rebellious generation, covered Apollo 11 for a book project, and found himself stirred and alarmed in equal measure. On the one hand he found the thunderous Saturn V immensely moving. "I had a poor moment of vertigo at the thought that man now had something with which to speak to God. A ship of flames was on its way to the moon."

Mailer even wondered if the tirelessly energetic WASPs were right in their ambitions for progress, and the hippies, feckless, lazy, and doomed? At the same time, he was depressed and frustrated by the Space Age's bland public face:

Everybody in NASA was courteous, and proud of their ability to serve interchangeably for one another, as if the real secret of their discipline and their strength and sense of morale was that they had depersonalized themselves. I had never before encountered as many people whose modest purr of efficiency apparently derived from being cogs in a machine. They spoke in a language not fit for a computer of events that might yet dislocate eternity.

This aerial view of the Johnson Space Center near Houston shows the scale of NASA's influences on earth as well as in space.

Mailer had the good grace to question his assumptions, but to some people Apollo seemed a cold and clinical achievement, a stark warning about the perils of a technological society. To them NASA was the chilly realm of white middle-aged men in uniform white shirts and dark suits; men made of flesh and blood, yet incapable of expressing emotion, rendered soulless by mechanistic conformity. The irony is that Apollo was almost entirely a product of emotion and instinct. Logic was far down the list of reasons for doing it. The machinery made possible the performance of certain tasks. It still remained up to humans to determine *what* those tasks should be. Let us stress, yet again, that America did not choose, single-handedly, to go the moon. There would have been no international prestige at stake unless, at some level at least, most of humanity shared the same dream. Rationalism handled the "how." Human emotions were always in charge of the "why."

John F. Kennedy attends a NASA briefing in 1962. NASA chief James Webb and Vice-President Lyndon Johnson are on the far left of the group.

James Webb: headstrong and ambitious, he was the right man at the right time.

THINKING BIG

Historians still debate the role played by President John F. Kennedy in shaping NASA's fortunes, but rather less attention is given to the man who actually led the agency while it prepared Apollo for the moon. It wasn't Wernher von Braun, despite his great celebrity and his skilled leadership of the Saturn V's development. The man in overall charge of America's space program was an unlikely figure, a burly, talkative bureaucrat with little knowledge of space or rockets. What *he* understood was politics, and how it intersected with business.

When President Kennedy took office in 1961 NASA was still a relatively minor arm of the federal government. The White House advisors made contact with at least seventeen distinguished scientists and business leaders, asking if they might like to head the agency, but at that time NASA didn't seem such a tempting prospect. In January 1961 it had sent a chimpanzee called "Ham" to the edge of space, riding on top of a slender rocket hastily adapted from a medium-range Army missile. Ham's mission, a preliminary towards launching astronauts on similar suborbital hops, lasted sixteen minutes and carried him a mere 420 miles from his original launch site on the east coast of Florida. There were some malfunctions during Ham's short ride. When he was plucked from the Atlantic after splashdown, frightened and wet, but safe, the new man in the White House, John F. Kennedy, wasn't that much impressed by the whole thing. His science advisor Jerome Weisner warned him, "A failure in our first attempt to place a man into space would create serious national embarrassment. We should stop advertising this as our major objective." No wonder no one wanted the NASA job. In some desperation Vice-President Lyndon Johnson strong-armed an unlikely candidate, a lawyer from North Carolina, into accepting the post.

James Edwin Webb was born in October 1906, in the small town of Tally Ho, North Carolina. As a young man, his first jobs were classic small-town means of helping his family to get along: working on farms, helping out in nickel-and-dime stores, and truck driving on construction sites. His father was the superindendent of schools in Granville County, and his mother was equally concerned about education and social progress. The highly intelligent young man cut his teeth as an education bureaucrat in the 1930s, and signed up to be a part of President Roosevelt's New Deal, a bold attempt to get America moving again after the Depression years, using vast sums of tax money to pay for infrastructure developments and social projects. Webb was hooked on this idealistic environment. He also turned out to be a smart political operator and a successful businessman. In 1938, by now happily married, he joined a Washington law firm, then became an executive at the Sperry Corporation in New York as it geared up for the coming war, before leaving in 1944 for a tour of duty with the Marine Corps.

> ## "*Somebody* in the government has got to tie it all together!"
>
> ### James Webb, 1961

After the war, Webb returned to public service, rising to the rank of Director of the Bureau of the Budget, a position he held until 1949. President Truman then asked him to serve as Under Secretary of State. But when the Truman administration ended in 1953, Webb didn't like the

idea of working for the new and staunchly Republican president, Dwight D. Eisenhower. He left Washington for a senior position at Kerr-McGee, the giant oil and nuclear power company in Oklahoma City. And there his contribution to the greater sweep of history might have come to an end. He could have finished his career as a talented and successful businessman just like many others, albeit with some flair for charitable works and public service: the embers from his youthful New Deal fires not yet burned out. Today, we might not have had cause to remember him in any broader context. But history wasn't finished with Jim Webb.

On January 17 1961, just as President Eisenhower prepared to hand over to a new leader with ideas very different from his own, he gave an extraordinary address to the nation, in which he reminded American citizens that a gigantic armaments industry had been created since the Second World War, where none had existed previously. In his opinion, it threatened to change all of society for the worse:

In the councils of government, we must guard against the acquisition of influence, whether sought or unsought, by the military–industrial complex. The potential for the disastrous rise of misplaced power exists and will persist. We must never let this endanger our liberties or our democratic processes.

It took just eight years for NASA to progress from the tiny Redstone rocket (below) to the immense Saturn V (left).

Eisenhower also warned that "public policy could become the captive of a scientific-technological élite." He would have been dismayed if he'd known that the newly-formed NASA would soon be under the control of a man who believed that, rather than being any threat to America's future, "a scientific–technological élite" could be the solution to all its ills. Webb believed that industry wasn't meeting the challenge posed by Soviet strength. He also felt that the colleges and universities were not creating a sufficient number of graduates properly equipped for the technological world that lay ahead. "You see, you've got the physicists here, the astronomers here, and the chemists over there— *somebody* in the government has got to tie it all together!" Webb said. And he was determined to be that somebody.

If President Kennedy had known how important space was about to become, he would undoubtedly have appointed someone more to his liking as chief of NASA. The political writer Charles Murray observes:

Stocky and voluble, Webb was of an older generation than most of the others in this new administration, and he came from a different world. Instead of Harvard and wire-rimmed glasses, clipped accents, and dry wit, he was University of North Carolina, all rumpled collars and down-home homilies: a good ol' boy with a law degree.

But he was the right man at the right time. While many NASA insiders thought that the great new lunar adventure announced by Kennedy in 1961 must surely continue for generations to come, Webb understood that this was a fragile, fleeting moment in history, a rare window of opportunity that would slam shut as suddenly as it had opened. Moving with a speed and decisiveness that seems hard to imagine today for any government bureaucrat, he swiftly consolidated NASA's influence across a vast crescent-shaped swathe of states: California, Texas, Mississippi, Louisiana, Alabama, Florida, Virginia, Maryland, and Ohio, the one inland state. Webb spent most of his time at NASA headquarters in Washington, from where he could keep an eye on the politicians.

A sensible way of running a rocket facility might be to put the launch pads, control centers, and construction hangars all in one location, and no one doubted Webb's reasons for expanding a small Air Force missile testing station at Cape Canaveral, on the Florida coast, into a giant moonport. But why build the mission control center so far away, in Houston, Texas of all places? And why build the Apollo capsules in California, booster components in Alabama and Louisiana, and yet more huge chunks of equipment in Long Island? His wide dispersal of NASA operations and private industrial contracts angered many people, not least the Congressional representatives of states where the NASA contracts were not so thick on the ground. They saw the distributions as politically motivated, perhaps even unethical. But Webb was almost impossible to intimidate. "I made up my mind early in the game that I couldn't let anybody dictate decisions that were at a technical level, whether it was the President, the Vice-President, or the scientists," he said in later years.

> "In the councils of government, we must guard against the acquisition of unwarranted influence by the military–industrial complex."
>
> **President Eisenhower, 1960**

SPREADING THE BENEFITS

He was playing a clever game. He wanted to ensure that as few Congressional representatives as possible would wish to block NASA's funding, for fear of losing space-related jobs in their particular home states. Hence, the spreading of NASA activities across the country. Second, as an old-style Democrat he believed passionately that one of the proper purposes of any large government undertaking should be to distribute jobs and education opportunities as widely as possible across the nation. While NASA's workforce grew to 30,000 people in the mid-1960s, private contractor staff amounted to ten times that number. It could never last, but for a few years at least, Webb's large-scale manipulations delivered jobs and a sense of self-worth and idealism to half a million space workers, from rocket scientists and computer experts to metal benders, building contractors, secretaries, and caterers. NASA's expansion triggered real-estate booms in Florida, California, Texas, Alabama, Virginia, and other states besides, as the various "field centers" grew from modest laboratories into complexes the size of small cities, and with suburbs to match.

Webb also fought behind the scenes to ensure that the Apollo project would not define NASA to the exclusion of everything else. The less glamorous robot probes and science projects were just as important, he believed. We take this multi-faceted character of space exploration for granted, but in the feverish atmosphere of the 1960s moon race, it had to be defended. President Kennedy made many eloquent speeches about the benefits of Apollo, and always steadfastly supported NASA in public, but privately he told Webb that the estimated costs of reaching the moon appalled him. "I'm not that interested in space," he admitted bluntly at a White House meeting in November 1962. "We're ready to spend reasonable amounts of money, but we're talking about these fantastic expenditures, which wreck our budget on all these other domestic programs, and the only justification for it, in my opinion, is because we hope to beat the Russians!"

Kennedy wanted an assurance: did Webb think that Apollo was NASA's number-one priority? "No sir, I do not," Webb replied. "I think it is *one* of the priorities." He argued that America would best benefit from a broader range of activities in space. Onlookers were amazed as a nasty shouting match threatened to brew between the two men, but Kennedy was apparently swayed and Webb's view prevailed. Throughout his tenure at NASA he fought to protect and promote the unmanned scientific and astronomy programs that we now celebrate as among the agency's most valuable achievements: probes to Venus, Mars, Jupiter, and worlds beyond, and the first steps towards funding an orbiting space telescope.

These are just the uppermost assemblies for the Saturn V's huge F-1 rocket engines.

Roosevelt had championed technocratic "Big Thinking" to salvage an ailing nation in the 1930s; wartime mobilization made it mandatory; Eisenhower had grave doubts in an era of 1950s post-war complacency but continued to sanction it; Kennedy took it pretty much for granted

as part of his New Frontier philosophy, and Lyndon Johnson tried to keep its dying embers alive for his Great Society, even as the war in Vietnam crippled his budgets and broke his spirit. Presidents Nixon and Carter eventually slammed on the brakes, but in the last decade before the demise of Big Thinking, NASA became one of its most committed champions. Jim Webb, Vice-President Lyndon Johnson and Robert Kerr, their ally on the Senate Space Committee, believed that the rocket's industrial impact on the ground was just as significant as anything it could achieve in space.

In those days, politicians were accustomed to the idea that nations could be run *efficiently*, if only the right system could be invented. The perfect society, led by selfless philosopher–kings, is one of the oldest dreams of politics, and it enthralled many governments in the last century, sometimes for the good, often for the bad. Today, in an age of free-market economics, we have forgotten that America's political élite once believed that their influence over the new forces of science and technology was crucial.

Was the technology of Apollo turning men into machines? This is the "firing room" at the launch complex.

Look at the advertising pages from America in 1961, and we see a country incandescent with material success. Behind the scenes, industry was coasting complacently along, producing outdated consumer goods: cars and refrigerators, lawn mowers, hair driers, and kitchenware. Optimistic marketing campaigns persuaded Americans to buy products like never before. The economy seemed strong, but the drive for short-term corporate profits couldn't satisfy the longer-term requirements of society as a whole. Perhaps even more seriously, private industry alone couldn't ensure technological supremacy on the military front: an overwhelming problem with Soviet Russia apparently at its zenith. Thinking Big was obviously the government's job. A massive federal expenditure on high-tech scientific research and development was initiated with the aim of funneling innovations into the private sector.

NASA in the 1960s flaunted the technocratic dream at its most self-assured: spending vast sums of taxpayers' money on science and machinery in order to improve education, stimulate flagging industries, and redefine the management structures and beliefs of large organizations across the country—and all with minimum interference from the public. James Webb's vision for Apollo was nothing short of a model for social engineering on a grand scale. It wasn't the moon so much as a transformation of the entire nation that he and his kind were after.

"I had never encountered as many people whose modest purr of efficiency apparently derived from being cogs in a machine. They spoke in a language not fit for a computer."

Norman Mailer, 1970

THE PERFECT ORGANIZATION

NASA campaigned throughout the 1960s for a new style of rationalistic "space-age management," aimed at transforming large-scale projects across the board, and not just in rocketry. New procedures in administration, human resources, organizational structures, and record-keeping were pioneered in order to keep the fantastically complex lunar project on track, and for a while they seemed to promise benefits for many other areas of national life, too.

Apollo was supposed to stimulate developments in industry, inspire schools and universities, and show the rest of the world—including Russia, of course—that Americans could govern their country better than anyone else. According to the Smithsonian Institution's chief space historian Roger Launius, "NASA thought it was possible to create the *perfect* organization. They talked about applying their techniques to other challenges, such as homelessness and poverty, welfare and education."

One of the reasons why we feel nostalgic for the era of Apollo is our instinct that the federal government could run at least some things better in those days, and the space program apparently proved it. Webb and his colleagues in NASA's senior management believed that people functioned best when they were granted responsibility for their own work. There was a tendency to reduce layers of middle-management so as to create an organizationally "flat" personnel structure in which junior staffers could talk directly to senior managers without fear or hindrance. Rank and seniority were not terribly important. In fact, many people dreaded promotion in case it took them away from work that interested them. At its best, NASA could boast the most successful management of complex engineering and manufacturing projects that the world had ever known. Even so, NASA's enemies feared that it was overstepping its bounds by a country mile. In 1963, an alarmed senator, William Proxmire, told Congress, "The space program is probably the most centralized government spending program in the United States. It concentrates, into the hands of a single agency, authority over an important sector of our economy. It could well be described as corporate socialism."

And Proxmire was a Democrat. Maybe he had a point. A very strange relationship had emerged between NASA and the private aerospace sector. It wasn't an expression of pure free-market economics, because the space agency intervened in the fortunes of private companies by placing substantial contracts at the taxpayers' expense, while those companies, in their turn, became dependent on the agency's patronage; and the relationship wasn't exactly socialist either, because the companies remained privately owned, and were therefore permitted to keep their profits. Yet it was a mirror image of Soviet Russia's command economy. Politicians set the agenda, determined budgets and decided what space vehicles they wished to see manufactured (albeit advised by NASA) and what new technologies they wished to encourage. The tens of thousands of employees in the NASA-funded aerospace factories were, by any sensible definition, workers for the state. If only three percent of a typical rocket company's trade was conducted in the private free market, how could it be otherwise?

The patchy illusion of free enterprise was preserved by the rules of competition. Companies vied for federal contracts by exhibiting cost efficiencies, technical prowess, and other fine virtues in their sales presentations. But it wasn't just the prospect of cash that drove leading aerospace manufacturers to compete, with rare and strident fury, for that greatest prize of all: the chance to build a ship to go to the moon. One renowned aviation designer, Harrison Storms, had made his name building well-armed fighter planes. Apollo, he said, offered him a chance, at last, to make a "wonderful flying machine that doesn't have a gun on it."

And in the end, the sums of money involved were not so terrifying as Proxmire made out. A simple comparison between the population census figures and NASA's expenditure during the 1960s tells us that it reached the moon for less than two dollars a month from each American citizen. Per day, it was the price of a stick of gum.

3 SPACESHIPS OF THE MIND

The interior of the command module:
a maze of switches, dials, and
electro-mechanical systems.

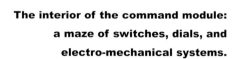

SPACESHIPS OF THE MIND

3

ONE MYTH WE CAN DISPENSE WITH is that the technology of Apollo is now so old-fashioned it can safely be consigned to the shelves of museums. On the contrary, a good deal of it still hasn't been matched. For instance, in late 2007 a team of NASA engineers designing new hardware for a new moonshot, planned for the coming decade, removed a panel from an old Apollo capsule on display in Washington DC's Air and Space Museum, and took a look inside the maze of wiring and pipework. They wanted the answer to a problem that had beaten the best minds in modern space engineering.

At the end of a mission, and just prior to re-entry into the earth's atmosphere, a crew capsule must separate cleanly from the equipment module that has contained most of its air, water, electrical power supplies, and rocket fuel. How is this to be done reliably, and in a fraction of a second? Dozens of power wires, data cables, and fluid conduits have to be disconnected in an instant, not to mention the three or four heavy attachment bolts that have held the capsule snug against the equipment module during the mission. The NASA team discovered inside the Apollo a tiny set of explosive guillotines, through which all the cables and lines snaked in a neat bundle. When the time came for the capsule to drop away, the guillotines sliced through metal and plastic as though they were butter. A backup set of guillotines, powered by a different electrical circuit, insured against failure—and all in a box the size of a car battery. Forty years later, Apollo still had lessons to teach.

In popular culture, spaceships are supposed to look sleek and futuristic, but even today the real-life hardware sometimes lacks a certain stylishness, and there is a subtle reason for this. Take a look, for instance, at how commercial airliners are built. The production runs are large enough to justify the initial costs of purpose-built factory tooling and templates. The cabins have moulded interior fittings that tuck neatly into the corners and meet seamlessly with the windows; the foldaway tables look like they belong on the back of the seats, and so on. There'll

The Apollo 11 command module on display at the National Air and Space Museum in Washington DC (right). The glossy exterior finish burned away during re-entry.

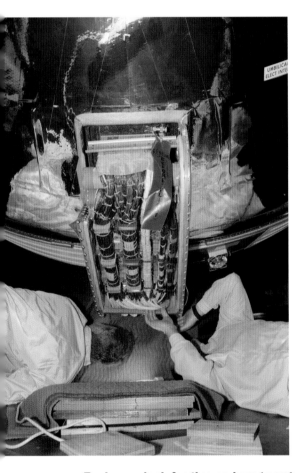

Engineers look for tips on how to get back to the moon (above) using rare unflown backup Apollo hardware (right) at NASA's Kennedy Space Center.

be some decorative flourishes, probably featuring the airline's logo, adding consistency to the various pieces. Inside and out, everything fits together into a more or less coherent whole. The same cannot be said of a typical spacecraft. These are hand-made machines, with populist styling absolutely not on anyone's priority list. It makes very little business sense for aerospace companies to create production lines for machines that are only going to be built a few times. Consequently there is a nuts-and-bolts feel to most spacecraft that causes some unprepared observers to think of them as disappointingly crude, almost wilfully antique.

The cultural reference point for what spaceships are supposed to look like is almost certainly Stanley Kubrick's 1968 science-fiction film *2001: A Space Odyssey*, itself a product of the Age of Apollo. Kubrick's prop designers first sat down at their drawing boards in 1964, at the start of what would turn out to be a long and complex production. The ships they invented still look futuristic. It is fascinating that the new space tourism companies, such as Virgin Galactic, have created interiors for their suborbital spacecraft that satisfy our expectations of how things *should* be, with digital control panels, soft padding on the cabin walls, and space meals you can suck through a straw, just like in *2001*. This is all an illusion, a marketing ploy. In real life, the interiors of space shuttles or Soyuz capsules look quite similar to the Apollo cabins, at least in their detailing of bolts, screws, and switches.

> "As a teenager, I thought that everything that happened before I was born was just ancient history. I didn't think I needed to know all about it. So I can't be too critical of this generation."
>
> **Neil Armstrong, 2001**

Visitors to aerospace museums sometimes expect gleaming science-fiction props. The real spacecraft that they encounter can leave them unmoved. People peer through the hatch of an Apollo capsule and see a dimly lit interior encrusted with ancient clockwork. There are no touch-sensitive screens or plasma displays. The instrument panels are a mechanical maze of switches and dials, and the thousand-fold electrical energies that once powered them are no longer in evidence to lend them some life. People are struck by how clunky, or even downright primitive, everything looks; and in countless historical reappraisals of Apollo, we are told repeatedly—aren't we?—that a child's little hand-held electronic game is more advanced than the computers used for Apollo.

Perhaps life changes so fast now that our culture has lost the ability to conceive of any world older than last year's. If we want to gain an impression of how futuristic Apollo seemed four decades ago, we need to make a couple of time-travel journeys of the imagination. Picture Louis Blériot in 1909 making what seemed at the time an epic journey, crossing the English Channel by air in a monoplane of such fragile design it seems unthinkable, today, that anyone could have entrusted their life to it. Look also at Lindbergh's plane, the *Spirit of St Louis*. Would you expect that machine, with its home-made finish, to have carried a man safely across the Atlantic Ocean? Yet when these two famous vehicles flew their missions, they were heralded as portents of a new world. They were *futuristic*.

John Knoll's digital recreation of the view from Armstrong's window as the lunar module approached close to a dangerous crater.

To the generation who witnessed it, Apollo was futuristic and then some. Imagine the command module brand new, completely covered in mirror-smooth silver foil insulation so that it looks like a single piece of polished metal. (The foil burns off during re-entry, leaving the capsule a drab brown color. It's best to think of it in its launch-day finery.) Now visualize the interior of the spacecraft, surgically clean and brightly lit so that the astronauts can see what they are doing. There is not a single scuff mark in the paintwork. The air-conditioning units are humming, and the control panel is a shimmer of lights and trembling dials. Apollo seems almost alive. By the standards of the 1960s this was the most advanced machine in history.

A CHILD'S TOY

The same applies to Apollo's on-board computer. True, the working memory inside it was only 256 kilobytes—and, yes, that capacity is easily outstripped by the cheapest modern toy. The computer on which this book was written pushes gigabyte files through microchips the size of stamps in less than the blink of an eye; but *this* computer is plugged into just a couple of fairly simple and low-powered interfaces with the outside world: a printer, a backup disc drive, and a telephone socket. By contrast, the input and output wiring of the Apollo computer connected it with a bewildering array of heavy-duty devices, from radio and radar systems to gyroscopic compasses and optical star trackers, and, ultimately, to the gas thrusters and powerful rocket engines that actually drove the ship through space. The computer responded to what the spacecraft was telling it, and told the craft what to do next. When the astronauts punched instructions into the computer's keypad, or nudged their pistol-grip thrust controllers to alter the spacecraft's orientation, these inputs were simply another of the myriad factors that the computer had to deal with and respond to, delivering its output in the form of tangible action, expressed as movements of the spacecraft itself.

"Our autopilot was taking us into a very large crater about the size of a football stadium, with steep slopes, covered with very large rocks. That was not the kind of place I wanted to try and make the first landing."

Neil Armstrong, 2005

A vast effort of design went into making sure that the computer could recover from crashes in less than half a second. Even when Apollo 12 was struck by lightning soon after launch, and the whole interior of the Apollo capsule blacked out for a moment, the computer survived. It was a reliable piece of equipment because it *had* to be. Astronauts' lives depended on it not failing. Forty years later, PC manufacturers and software writers have simply not delivered to us that same robustness. Our computers are vulnerable to the smallest upsets. Until one sits on your desk that can work on many complex calculations for a fortnight without seizing up, no one can talk again of the primitive nature of Apollo's computer.

Contrary to any number of accounts, not least those of some of the astronauts themselves, there was not one moment during any lunar mission (apart from Apollo 13) when humans had to "take over" from wayward electronics and fly the vehicles themselves for fear that some misfiring circuit was steering them to their doom. A famous Apollo 11 incident is worth retelling, because—contrary to the way that it has been interpreted in some popular accounts—it is a story of men and machines working in harmony, rather than against each other, in what's now commonly known as a "fly-by-wire" system:

Just as the lunar module (LM) was plunging towards the lunar surface in final approach, the computer display in the cabin suddenly started flashing. "Program alarm," said Armstrong. Alongside him, Buzz Aldrin punched a key on the display board. "1201," he reported.

At Mission Control, 26-year-old Steve Bales was sitting at the guidance console, monitoring the LM's computer system. Senior flight controller Gene Kranz came on the microphone loop and demanded to know, "What's a 1201?" Bales was suddenly in the spotlight. The LM was rushing towards the moon and the computer was saying that something was wrong. He needed a few seconds to think. "Stand by," he replied, trying to buy time. But Aldrin wanted an answer right away. "Give us the reading on the 1202 alarm," he said. In astronaut-speak, he was asking if Armstrong and he should abort the landing. Bales had no time left. The LM was plunging into what NASA insiders privately called the "Dead Man's Zone."

After undocking from the orbiting command module to begin its separate voyage, the LM's crew module (ascent stage) could punch away from the lower descent stage at more or less any time if anything went wrong. The ascent stage's engine could power the astronauts back into orbit to rejoin the command module. But the Dead Man's Zone was a brief period of uncertainty that nobody could do anything about. Within the last three minutes of the landing approach, there was a ten-second phase when the LM was hurtling downwards so fast that if the ascent stage's engine fired in an emergency abort, the entire fuel reserve would be wasted counteracting the downward momentum. The ascent stage wouldn't be able to climb back up to join the mothership. Instead it would crash on the moon.

As the computer alarms sounded, the Dead Man's Zone was coming up in twenty seconds. If Bales was going to recommend an abort, he had to do it right now. Making the bravest and possibly the most reckless decision of his life, he spoke into his headset for Kranz and all the other controllers to hear. On the tapes that survive to this day, his voice is shaky, fearful. For this moment at least, the entire Apollo program rests on his young shoulders, not to mention the lives of two men out in space. "We're 'Go' on that alarm," he said. This was a signal for the astronauts to ignore the alarm and carry on.

> "The lunar module may have seemed terrifyingly fragile, but we had faith in the people who built it. We entrusted our lives to it."
>
> **Buzz Aldrin, 1997**

Kranz was surprised, but he trusted all his controllers completely. He let the decision ride, and the LM plunged down into the Dead Man's Zone. Strictly speaking, Bales should have called off the landing, but he didn't quite believe the 1201 code. Even as the computer was flashing warning signals, it was still feeding the proper information to the astronauts' control panels, and the descent was going pretty much according to plan. The only solid evidence that Bales had was his memory of similar alarms cropping up during training sessions. Forty years later, and Bales still isn't sure about the decision he made in that terrifying moment. "Nobody really knew what was causing the problem, and I couldn't be a hundred percent sure that the judgement I was making was okay. It was based more on instinct than hard facts." Bales decided that so

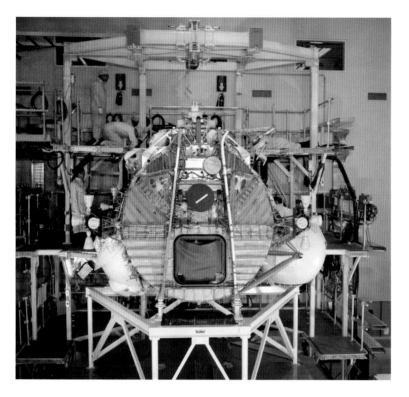

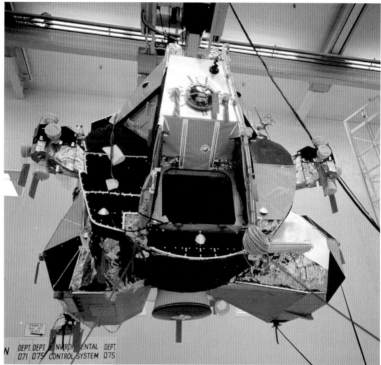

Assembly of the lunar module at the Grumman factory in Bethpage, New York, and (overleaf) John Ortmann's digital recreation of the two Apollo spacecraft.

COLUMBIA

EAGLE

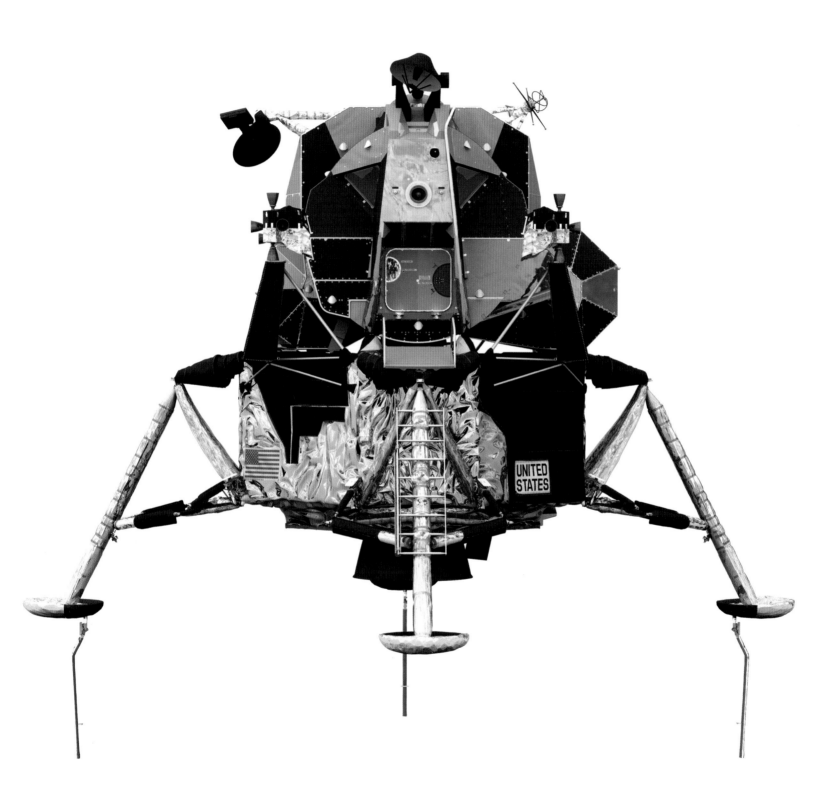

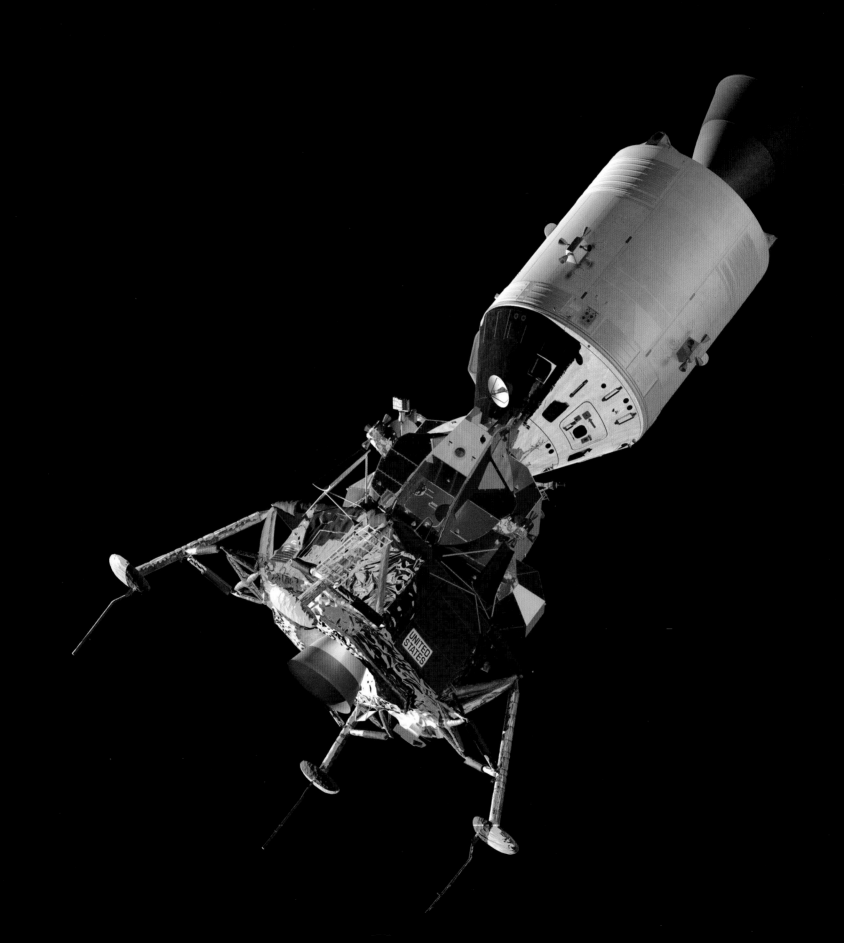

The command module is prepared for mating with the Saturn V (above). Blue plastic sheeting protects the brand-new capsule from damage.

John Ortmann's artwork (left) shows something we never see in the mission photos: both spacecraft docked.

long as the computer was still working, it was working, and that was that. Just as he was recovering from the computer scare, CapCom Charlie Duke was radioing a warning up to the Apollo crew: "Sixty seconds." Listening to the original NASA dialogue tapes, it's impossible to tell that anything was wrong, but in technical shorthand, Duke was really saying to Armstrong, "Your engine only has sixty seconds-worth of fuel left in its tank."

Duke called out another warning. "Thirty seconds." Chief astronaut Deke Slayton, standing behind Duke, said softly to him, "Shut up Charlie, and let 'em land." Any moment now, and the LM's engine was going to sputter and die. The instant it did so, the ascent stage would automatically separate and begin an emergency return to lunar orbit. Firing the ascent stage under these chaotic circumstances was no guarantee of safety, even out of the Dead Man's Zone, but it was better than simply crashing the LM onto the moon.

Kranz and his team then saw something amazing on their telemetry screens. With less than a hundred feet remaining before touchdown, it seemed that the LM was pitching forward and skimming over the lunar terrain at thirty miles per hour. Had the two best pilots in NASA suddenly lost control of their ship? Another ten seconds crawled by. At last, Aldrin radioed, "Contact light. Mode control to 'Auto.' Engine Arm off." Those were, in fact, the first words ever spoken by a human being on the surface of another world. Amstrong chipped in a moment later with phrases more easily understandable to the general public. "Tranquility Base here. The Eagle has landed."

Mission Control erupted into applause, and Duke's radioed response, "We're breathing again. You've got a bunch of guys about to turn blue. Thanks a lot," was almost lost among the whoops and hollers of delight. Amstrong sounded apologetic when he radioed to explain what had happened in the last minute of descent. "Houston, that may have seemed like a very long final phase, but the [guidance system] was taking us right into a crater with a large number of boulders and rocks." He had needed a few precious extra seconds to hover a few meters above the surface and nudge the ship forward until he could find a safe place to land. Later, the newspaper and television reporters talked of Armstrong heroically "switching off the computer" and "seizing manual control" to steer the little craft away from danger. In fact there was never any conflict with the computer, because the astronauts were always supposed to be able to choose the exact patch of ground they wanted to land on. That's why the LM had windows for them to look out of.

Apollo 11's landing site on the Sea of Tranquility had been chosen on the basis of the most scrupulous orbital scans and telescope observations, but matters were very different when the LM was hovering just a few tens of feet from touchdown. At that point, dangerous rocks, boulders, and craters, too small to be observed from orbit, made themselves apparent. Armstrong quite properly took evasive action to avoid hazards, working in tandem with the computer—not against it—to bring the little craft down safely. Like an upright pencil teetering on a fingertip, the LM was balanced on a plume of engine thrust from a single nozzle, and it was only with constant help from the computer that any human could maneuver the craft. As for the apparently urgent alarms flashing on the display panel: they did not foretell impending failure. They were just warnings that several computational tasks were baying for attention all at once and, given the urgency of the landing, the computer was sending non-essential jobs to the back of the line. A late decision had been made, prior to the mission, that it might be sensible to have one of the LM's radars keeping tabs on the orbiting mothership, while the other dish pointed down to the lunar surface. The computer decided that the mothership data had to wait. But it did *not* misfunction. No wonder its young designers at the Massachusetts Institute of Technology (MIT) felt that their achievements had been somewhat misrepresented by the media. And in their old age, how weary they must have become of hearing that you could navigate to the moon today with a child's toy.

Apollo 11 command module pilot Michael Collins strikes a heroic poise inside the training simulator.

THE LANGUAGE OF APOLLO

The endless repetition by the media, parrot-style, of the NASA jargon soon bored the public, for this was not a language in which ordinary people would expect to speak about such epic events. Soon the events themselves seemed less momentous: just an inevitable product of a Lunar Orbit Insertion and an Extra-Vehicular Activity followed at the Time-Nominal moment by a Trans-Earth Injection Burn. Many journalists and TV reporters were diverted from the true drama by this barrier of language, occasionally steeping themselves in it like "wannabe" space travellers, but more often feeling dislocated by it. As Norman Mailer confessed:

In fact I was bored. Sitting in the Manned Spacecraft Center movie theatre, I noticed the press reporters were also bored. We all knew the engine burns would succeed and Apollo 11 would go into the proper orbit. There seemed no question of failure. I could not forgive the astronauts their resolute avoidance of an heroic posture.

It is only now, some forty years later, that we are gradually learning to appreciate just what a "heroic posture" the men of Apollo actually adopted, each and every time they climbed into those capsules.

NASA's acronyms are like medical terms, precise, exactly defined, specialized, and devoid of emotional loading. A potentially terrifying fuel leak on the pad is an "anomalous discharge." A cataclysmic explosion on a shuttle is "a major malfunction." No one dies or is killed. Rather, a crew "fails to respond." Charlie Duke didn't radio to Armstrong, "According to my readouts down here in Mission Control, you have maybe a minute or so of fuel left. Is that right? Is that what it says on your panel, too? Anyway, you be careful not to crash, you hear?" He simply said, "Sixty seconds." Space people share with the military an instinct that distressing situations can only be handled by keeping some sense of emotional distance. Keeping calm is important for men and women operating dangerous machinery. Astronauts cannot afford the ambiguities of poetry. Their lives often depend on knowing, in a split second, exactly what some particular word or phrase means.

Hundreds of thousands of people are still fascinated by space exploration today, and millions more take at least a passing interest, but those multitudes may not be sufficient to ensure a long-term continuation of the human space flight at a major national level. The drama of human rocket flight is as great now as it ever was, but astronauts do not seem to captivate us as they once could, and rockets have lost some of their ability to inspire awe. There is a problem in the language of space if young people no longer find as much excitement in the adventure as their parents did fifty years ago, when they, in their turn, were young. The cockpit chatter of space professionals will always remain terse and business-like, but perhaps NASA needs to speak to the rest of us in a more honest way that engages our emotions as well as our minds.

Armstrong in the lunar module simulator: a machine even more complicated than the real thing, because of the many exterior computers, photo projectors, and other systems that made it work.

"Machines are getting better all the time, but there's still a place for us—some reason for us, *Homo Sapiens*, to carry on existing."

Neil Armstrong, 2001

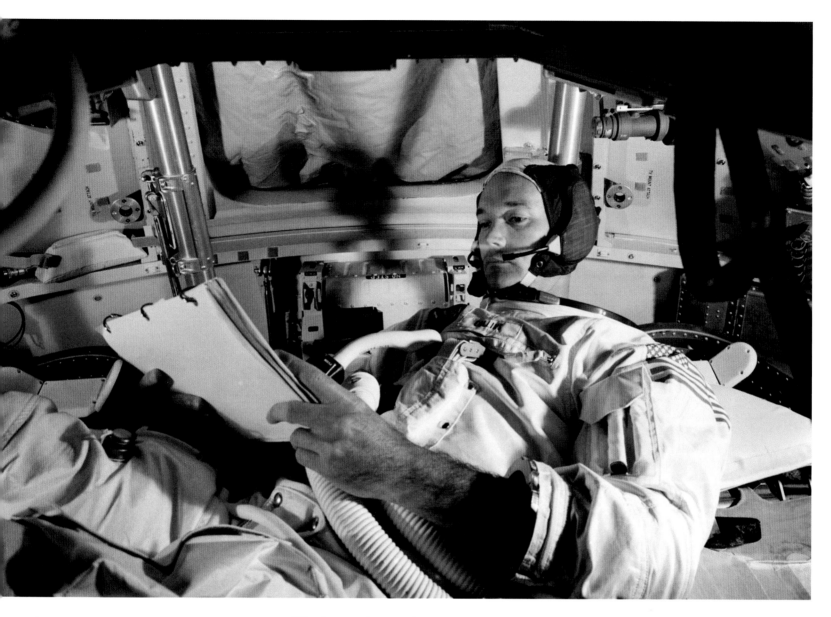

Collins in the command module simulator. Some sessions would last for several days on end, in order to acclimatize the astronauts to long periods inside the capsule.

We also need to send more people into space with more of a gift for telling us about what they see up there. By all means, let cool-headed minds prevail over the cockpit controls, but give the dreamers a chance, now and again, to stare out of the windows and ponder. Apollo 11's Michael Collins recognized this when he said, "I think a future flight should include a poet, a priest, and a philosopher. Then we might get a much better idea of what we saw."

THE ART OF SPACE

Artworks other than photography should also play a role in space flight. In 1962 Hereward Cooke of the National Gallery of Art in Washington DC wrote a letter of invitation to a number of prominent artists, inviting them to tour NASA and create works based on their impressions. He was eloquent about the need for both art and science in any space endeavor. "When a major rocket launch takes place, more than two hundred cameras record every split second of the activity," he wrote:

Every nut, bolt, and miniaturized electronic device is photographed from every angle. But the camera sees everything and understands nothing. It is the emotional impact, the interpretation and hidden significance of these events that lie within the scope of the artist's vision.

Cooke assured the selected artists that they would be given access to NASA facilities, and would be subjected to no editorial pressures. The only stipulation was that "the materials used must be of proven permanence. I do not want fading pictures in a government archive."

Paul Calle, famous for his scenes of life in the Old West, took up the challenge, making superb pencil drawings of astronauts and their capsules. Robert Rauschenberg's abstract silk-screen prints seemed both celebratory and mildly sarcastic at the same time; Lamarr Dodd captured an impressionistic morass of wires, switches, and dials, with silver-clad humans embedded into the machinery of their ships. James Wyeth's watercolors showed the forlorn scrublands surrounding NASA's launch complex. Rockets had to keep their distance from the ordinary human realm of towns and streets and family backyards. The drab safety-zone territories around the pads were a sort of endless "nowhere." A more inward landscape was explored by Mitchell Jamieson: the psychological drama behind the space age's political rhetoric. It needed only the lightest sweep of an artist's brush to expose the religious and mystical desires implicit in cosmic exploration. Jamieson's geometrically fragmented astronaut in his spacesuit brings to mind the saintly figures in a cathedral's stained-glass window.

Astronauts in training have a forlorn and vulnerable look to them. They are so out of place on the earth.

Aldrin, in the red jacket, gets to grips with a Hasselblad camera while Armstrong looks on.

The resulting art collection, a major body of work, is now under the guardianship of the Air and Space Museum in Washington DC. However, times have changed. In 2004 the singer and performance artist Laurie Anderson accepted a year-long commission as artist-in-residence for NASA. With her penchant for dreamlike electronic experimentation and her restless curiosity about modern technological culture, she was the perfect artist to take a quirky, sideways look at the space business. Sadly, by then Congress had forgotten the importance of art. When it got wind of Anderson's modest honorarium of $20,000 for a year's work, there were calls to halt her project, and unfortunately they were succcessful. One Congressional representative said that "NASA should not be spending taxpayer dollars on a performance artist."

That narrow-minded kind of thinking could cost America its future in space. Just think: if those medieval cathedral-builders had tried to save a few scraps of money by cutting back on the painted altarpieces, or the filigree stone carvings on the parapets, then so many of the intricate and expressive details that have helped those cathedrals to speak to future generations would have been lost to us, and we might not be so keen to preserve them, even at some expense. In much the same way, NASA should be allowed to communicate through art as well as rocket engineering, so as to reach as wide a spectrum of society as possible.

4 AN ELEMENT OF RISK

AN ELEMENT OF RISK

4

THE UNAVOIDABLE GULF between astronaut-speak and the way that rest of the world talks remains a problem today, for people are still surprised when missions occasionally go wrong. This generation is not accustomed to danger, either personal or global. No language has yet been invented that can persuade the public to fund rocket adventures in which, at some point, people are bound to die. They expect that NASA should be able to predict all risks, and that space vehicles fuelled with the energy equivalent of small thermonuclear bombs should be made as safe to fly as ordinary aircraft; at least, they should be if the government is paying for them with tax dollars. Surely all that dry, technical babble means that NASA must understand what it's doing by now, and that failures must therefore be the result of incompetence?

But rockets can be lethal, and space will always be an environment hostile to life. People make errors of judgement, and they can be very serious ones that lead to disaster; and sometimes people's animal instincts save a mission even when all the dials and read-outs are indicating trouble. Apollo 11's little lander could so easily have crashed on the moon, plunging America into a fantastic international humiliation that would have echoed to this day. The doomed space shuttle Columbia could so easily *not* have been struck by fuel tank debris in January 2003.

To date, fewer than six hundred people have flown into space. Three were killed before their craft ever left the earth, another crashed to the ground, three were suffocated during re-entry, seven were blown apart soon after launch, and seven died in an atmospheric disintegration of their ship. This is a fatality rate of one space traveller in twenty-eight. A dangerous profession indeed. Yet somehow, the illusion persists that journeys into space should be routine, and that accidents are blameworthy. As a consequence, it is harder, today, to build a ship for the moon than it was forty years ago. No matter how much money NASA spends, and how many checks it makes, it can never satisfy the modern craving for perfect safety. Astronauts in the Apollo era accepted the risks because the prize on offer seemed worthwhile to them: the chance to push

Armstrong's official portrait prior to Apollo 11, and (below) his younger self in the early 1960s, not long after he had joined NASA.

Neil Armstrong as he is today: shy, unwilling to promote himself, and refusing to adopt "a heroic posture."

at the frontiers of human experience and gain at least some degree of public glory for flying higher, faster, further away than anyone else in history. If the public understands that something new is being accomplished with a space mission, it tends to be more understanding about the potential hazards.

Part of the mystery of Apollo is that our generation is not so used to taking risks. When the bid for the moon was announced in 1961, almost all American leaders or managers in the prime of their professional adult lives—say, 40 to 50 years old—had fresh memories of the Second World War, and for them the Cold War was business as usual but with bigger and scarier bombs. The idea of risking the necks of a handful of astronauts in the service of their country didn't seem all that unusual. Apollo was accepted as a dangerous adventure. Indeed, that was part of its appeal. NASA's astronauts, and their rivals in the Soviet system, accepted huge personal risks in the struggle to reach the moon. In January 1967 three of them, Gus Grissom, Ed White, and Roger Chaffee, died in the most horrific circumstances imaginable, sealed inside a cramped capsule, a prototype for the Apollo, while it exploded into flame.

Some of the blame for the fire was laid at the door of North American Aviation, the California-based company responsible for building the capsule. Faulty wiring caused a tiny short-circuit spark. Ordinarily that should not have been a disaster, but NASA had chosen to pump pure oxygen into the Apollo's cabin for the astronauts to breathe. At the time of the fire, the oxygen supply was pressurized at higher than normal levels so that the capsule would not be contaminated by dust and moisture creeping in from the Floridian air. Oxygen is highly flammable under pressure, and somehow NASA missed this danger.

DOUBLE STANDARDS?

Congress expressed its dismay and blamed the fire on serious incompetence within NASA. Investigators concluded that many fire risks had been ignored or overlooked, such was the mood of confidence at that time after the brilliant successes of all the earlier flights. NASA had completed sixteen manned missions in two different kinds of spacecraft (Mercury and Gemini) without losing any crewmen. Just three months later, a Russian pilot, Vladimir Komarov, died when his parachutes failed and his Soyuz thudded into the Russian steppes like an unrestrained meteorite. The early years of space flight delivered what we would call a "steep learning curve."

In 1967 no one commented on the high standards expected of NASA in the same year that 56,000 American soldiers were wounded in action and 9,400 killed, during a campaign whose management was uncertain, whose purpose was undefined, and whose execution, even by the most charitable judgement, was flawed. The soldiers in Vietnam had little or no choice about their exposure to risks. The war could never have withstood the scrutiny that was applied to Apollo, but NASA's crewmen knew they were in a dangerous adventure and climbed willingly into their capsules. Some of them even expressed the view that they had it easy in comparison to their many aviator friends who were still in active military service. In April 1970, Jim Lovell, Fred Haise, and Jack Swigert, the crew of Apollo 13, nearly lost their lives when an oxygen tank exploded on the outward trip to the moon. There were many people in the space business who were rather relieved when the lunar landing missions were brought to an end in 1972 before a crew was actually lost out there. Senior Apollo manager George Mueller resented the fact that NASA had been under pressure to make unrealistic promises about safety:

How do you explain to the public at large that there's a certain danger in space flight, and you've got to accept that risk? We haven't been able to do that in our society. People are quite willing to accept risk for race drivers, for example, and clearly that's a hazardous operation. But when one of them gets killed, you don't have a Congressional investigation or a Presidential commission looking at it.

Michael Collins (above) showed flair as a writer. This helped him come to terms with the stresses of Apollo 11. His crewmate Buzz Aldrin (right) was a man driven to succeed, sometimes at the cost of his happiness.

"Space is not just about having fun.
It's also the pursuit of achievement,
of service, of challenge."

Buzz Aldrin, 1999

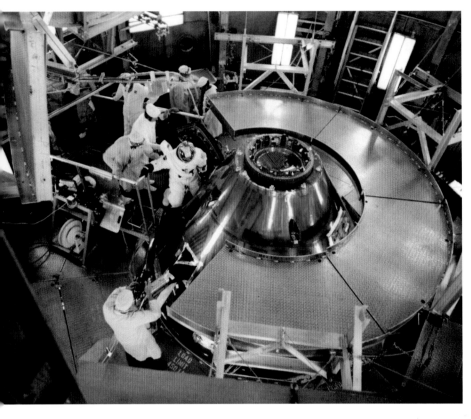

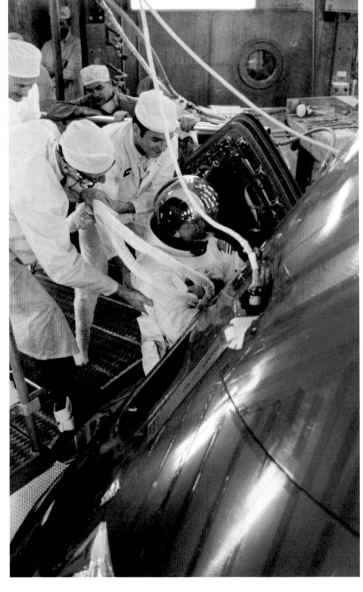

On November 9 1967, the world's largest and most powerful rocket, the Saturn V, lifted off for the first time, in an unmanned test mission known as Apollo 4. NASA announced that Apollo was back on track after the fire, at least as far as the launch vehicle was concerned. Unfortunately, the Saturn still had some nasty surprises up its sleeve. On April 4 the next year, another unmanned Saturn took off, designated Apollo 6. (The earlier Apollo 5 flight of January 22 was an unmanned capsule test using a much smaller rocket.) From the press stands five miles distant from the pad, everything looked wonderful. The five thunderous F-1 engines lit up and the Saturn rose into the sky on a gigantic column of fire. As it turned out, few newspapers or TV stations were concerned to report the launch, because a tragedy of global significance had turned their attention elsewhere. Civil rights leader Martin Luther King was shot and killed that same day. If the press pack hadn't been distracted by the assassination, they might have asked a few questions about that second Saturn flight.

For a start, two of the five F-1 main engines had failed shortly after lift-off, leaving the rocket unstable. Mission controllers kept their fingers hovering over their abort buttons, because at one point—too high in the trajectory for ground observers to notice—the rocket keeled over and headed nose-down towards earth before levelling out again. Somehow, the rocket's guidance computers compensated for the missing engines, and the upper stage, with its unmanned Apollo

"It would have been impossible for them to realize that the day might come when Americans would say, 'Oh yes—now, which one was he?'"

Tom Wolfe, 1979

Would you choose to spend a day, let alone a week or more, sealed into one of these capsules? Here, an Apollo crew tests the capsule in a vacuum chamber that creates conditions similar to those in space.

capsule, just about limped to orbit. It wasn't exactly dignified. After careening like a drunken elephant, the Saturn's upper stage ended up flying backward around the earth. The Apollo capsule suffered heavy vibrations during the terrible launch, and many of its control systems failed. Also, the conical section of rocket skin beneath the capsule, designed to house the spider-like lunar landing craft in future missions, had lost one of its four panels, leaving the Apollo, still perched on the Saturn's upper stage, structurally weak. The next task in the mission was to relight the upper stage's engine and see if it could boost the shaken Apollo in the moon's general direction. The engine wouldn't even start.

The mission was a catastrophe. If there had been astronauts aboard, they would probably have lost their lives. At best, they would have had to fire the small escape rockets on their capsule and make a dangerous bid for an emergency splashdown. The Saturn V wasn't safe to fly. In the light of all this bad news, a senior Apollo manager, George Low, made an incredible decision. Next time the Saturn lifted off, it would carry a full crew—and it would take them all the way to the moon. When NASA chief James Webb heard about Low's plan, he exploded in fury. "Are you out of your mind? You're putting our agency and the whole Apollo project at risk!"

THE RACE ACCELERATES

Webb had good sense on his side, but data from the Pentagon's spy networks soon persuaded him that Low's plan was worth the risk after all. There were only twenty months left before John F. Kennedy's "before this decade is out" deadline for a lunar landing ran out, but NASA wasn't just racing to redeem a pledge of honor made by a dead president. Soviet Russia was also heading for the moon. The intelligence agencies knew that the Russians' massive N-1 lunar landing booster was years behind schedule, and no great threat to the Apollo timetable, but a smaller Russian rocket presented more of a challenge. The Proton, still in use today, was at an advanced level of development by 1968. And so was the new multi-crew space capsule, Soyuz, which would become the sturdy Russian workhorse.

There was a reasonable chance that the Proton could send a "stretched-limo" version of Soyuz around the moon by the end of 1968, even if it had to be so crammed with fuel and oxygen that only one cosmonaut could fit inside. If this mission succeeded, then it would be another Communist triumph: first satellite in space, first animal, first man, first spacewalker, first woman, and now, the prospect of being first to the moon. If Russia did manage this makeshift trip in a Soyuz–Proton combination, then Apollo would be shamed into second place. Of course NASA would eventually make an actual landing on the lunar surface, but the world would perceive America to have lost in space yet again. There could be no further delays in Apollo. There was no choice but to launch the Saturn V on a lunar mission as soon as possible. The lunar module landing craft (LM) wasn't yet ready to make a flight, and the only short-term option available was to send Apollo 8's crew to the moon without their lander. First, however, someone had to find out why the Saturn V wasn't working, and fix it.

Thousands of engineers worked on fixing the Saturn. Analyzing the failures was tricky because there was no wreckage from the last launch to examine (unless you counted the expended first and second stages lying at the bottom of the Atlantic, beyond salvage). The only evidence was radio telemetry data describing the flight in numbers and symbols on computer print-outs. From this slender evidence, an incredible detective effort resolved the problem. Kerosene and oxygen fuel thrumming at high speed through the pipework of the five main engines had set up a resonance that shattered the fuel lines. It's a bit like the effect you get in a tiled bathroom when you whistle at a certain pitch and the whole room suddenly hums. Make the tiles out of several different materials at random, and you can eliminate the humming. With a similar theory in mind, the pipework for the F-1 engines was redesigned with a new mix of materials, and all the vibrations in the fuel lines were eliminated. It was a brilliant effort: probably one of the cleverest and least-known engineering achievements in history.

Collins looks warily at the complicated hatch mechanism in the nose of the command module.

Ground tests confirmed that the fix had worked, but it was still a huge risk to launch the next Saturn V with men aboard. However, it seemed necessary at the time. On September 14 1968, Russia finally succeeded in swinging an unmanned capsule around the moon and returning it to earth. In response, NASA sent Apollo 8 around the moon during the Christmas of 1968. That was what you call taking a chance. Men had ventured to the moon in urgent circumstances, flying new and hazardous machines, but somehow the true significance of that fact was not reflected by NASA's dry presentation. The dangers of the mission—precisely those dangers that might have excited and moved the general public into understanding just how incredible this achievement really was—were somehow lost in translation. This was why Tom Wolfe was able to write such prescient words about the astronauts in his 1979 book *The Right Stuff*:

The mantle of Cold Warrior of the heavens had been placed on their shoulders without their asking for it; and now it would be taken away again without their knowing that either. It would have been impossible for them to realize that the day might come when Americans would hear their names and say, "Oh yes—now, which one was he?"

THEY REALLY DID LAND

But even Wolfe, with his taste for a sarcastic aside, wouldn't have imagined that the day might come when the Apollo astronauts were accused of *faking* their missions. Many thousands of people around the world believe that we never landed men on the moon. They claim that the ghosty TV pictures of Neil Armstrong stepping onto the lunar surface in July 1969 were set up in a giant studio as part of one of the biggest conspiracies of all time. The simplest argument that the conspiracy theorists use is that there should only be one very concentrated source of light in the Apollo lunar photos—the sun—but many of the shots look as if they were snapped using some kind of much broader and softer illumination. The harsh sunlight on the moon should create deep black shadows, with no subtle shading in between; yet in most of the shots the astronauts seem suprisingly softly lit, like male models flaunting the latest craze in clumsy clothing. The shadows look as if they've been filled in by lights other than the sun.

The explanation is simple. The astronauts and their lunar module are standing on a gigantic gray-white photographic reflector. The moon's surface is a bright gray-white, as any pair of lovers will notice when they look up and see how brightly the moon bounces sunlight back into space. Photographers and movie makers are familiar with using large reflector panels to "fill in" the shadow areas of their subjects, particularly when counterbalancing very bright directional sunlight. On the moon, the surface terrain works a treat, and the astronauts are lit almost as perfectly as the NASA publicists could have hoped for.

Not to be confused with the real thing: Armstrong and Aldrin train for the moon walk in front of a mock-up lunar module.

The other argument from the conspiracy addicts is that countless stars should be visible in the black and airless lunar sky, yet there is not so much as a single speck of starlight visible on any of the 600 film frames snapped by Apollo 11's astronauts. This is a mystery that almost any photographer can solve after a moment's thought. An astronaut in a white spacesuit in bright sunlight needs an exposure of only a few fractions of a second, but stars are too distant and dim to register on photographic film unless exposure times of several seconds are used. This also explains why modern space shuttles and space stations photographed in orbit invariably appear against the same pure black sky. Time and again, space travellers tell of the vast panoply of stars they can see, yet those stars never appear on their photos.

Collins in a dummy command module cockpit, awaiting a centrifuge session.

A HISTORICAL ANOMALY

These are just the technical arguments against an ungenerous accusation of fakery. On a wider level though, the conspiracy stories probably reflect unease about Apollo's place in the history of machines. In 1886 Karl Benz patented the first automobile. Forty years later, the American car industry was manufacturing four million of them a year. The Wright Brothers flew the first human-carrying powered aircraft in 1903. Forty years later, commercial air travel and its darker cousin, aerial combat, were routine, and some people even had their own private planes. In the 1950s, a few large companies and institutions were using computers to do their payroll or solve complex equations. Forty years later, we all had one to play games on. And so it goes, and so it goes. The lunar adventure nourished countless areas of the economy in subtle yet important ways, speeding up developments in software, metallurgy, precision welding, and other profitable but unglamorous skills, such as personnel recruitment and management, risk analysis and information flow. But these benefits, although many and varied, haven't been so easy for most of us to identify. The more obvious impact that we might have expected, judged according to previous historical trends in machinery, are absent. We don't all have rockets in our garages and we can't all travel to the moon. This bucks the trend of just about any other branch of technology that has ever been invented.

Strangest of all, NASA and the US government tell us that it's almost as hard, and nearly as expensive, to build a lunar spacecraft now as it was forty years ago. And they're promising to do it in a dozen years or so, when last time it took less than a decade. Naturally, some people mutter that if we could accomplish Apollo using 1960s technology, then we should be able to

Three ordinary men. Collins once remarked that the main qualification for flying to the moon was to have been born at the right time—in 1930, give or take two or three years.

Armstrong narrowly avoided getting killed in this machine, a jet-powered training vehicle designed to mimic the behavior of the lunar module during its final descent to touchdown.

do it all again today, and at less cost, what with all the modern microchips and materials at our disposal, not to mention the vast backlog of knowledge and experience already accumulated in NASA's brain bank. The fact that we can't suggests that Apollo must have been some kind of a miracle the first time around. In a sense, that's just what it was. An accidental series of alignments in history made Apollo possible, bringing it to the forefront somewhat ahead of time. In the 1960s we weren't quite ready for lunar travel, yet a moonship appeared.

By the end of the Apollo era there was a wide sense of public distrust towards large-scale technological enterprises, but the counter-culture campaigners and peace-promoting liberals who so much hated the Vietnam war, and who feared the technocracy represented by Apollo, were *not* the people who assumed the greater measure of power in America after 1968. The true pull back of NASA-style Big Thinking wasn't just a liberal backlash against technology. It was the Republicans' response to what they perceived as the over-reaching intrusion of the government into the nation's affairs. Even as Apollo 8 was prepared for its circumlunar mission, James Webb stepped down as NASA's chief, exhausted and out of step with the changing mood, both in Congress and at the White House.

> "We didn't devote a great deal of our time to investigating the scariest aspects of our flight. Sometimes there is a morbid human curiosity about death-producing events."
>
> **Buzz Aldrin, 1998**

In 1969 Webb's successor, Tom Paine, found that President Nixon was, at best, lukewarm towards space, and Paine wasn't ready to accept that NASA had to change. He thought that the lunar landing missions had set a precedent for further large-scale space projects, but he was wrong. Apollo had been the exception, not the rule. It was Paine who was actually forced to call a halt to the moon program, cancelling missions 18, 19, and 20, and trying to save money and appease the new president. However, he did successfully salvage a working Saturn V for use in the last vestige of the Apollo program, the Skylab orbital workshop of 1973, America's first space station. Unfortunately, Paine was no match for Nixon's advisors, and he resigned after a tenure of less than two years. "I finally left because I did not think that I could deliver the kind of relationship with the President that the head of NASA really ought to deliver."

A SENATOR'S JUDGEMENT

On July 19 1979, a decade after the Apollo 11 lunar landing, the highly respected Democrat senator Margaret Chase Smith wrote an impassioned letter to be lodged with the Library of Congress. It stands as a fine testimony to Apollo. It is also a rebuke, on permanent record, for a nation whose gigantic achievements in space exploration were too carelessly cast aside:

I am saddened by, and disappointed with, three matters. The first is the public's apathy about the space program. Once we had a man on the moon, the American public lost interest. My second disappointment was the inevitable consequence of public apathy. It was the decline of the Senate Space Committee. When the public lost interest, its elected representatives in Congress found the space program less productive in terms of providing votes for reelection. After 1973 the Senate struggled to get someone to chair the Committee, because none of the sitting members were particularly interested. My third disappointment has been the lack of recognition for the person who put a man on the moon: James Webb, NASA's Administrator during its most crucial years. He had to take the heat and fire of partisan political attacks from headline-hungry politicians. I saw this first-hand in my work on the Space Committee! What recognition or material benefits did Jim Webb, and the thousands of people behind the scenes that he typified, receive? Minimal, if any—and today, they are forgotten people of a forgotten program. What we need today on the energy crisis is another Jim Webb.

The "energy crisis" Smith was referring to was the oil price escalation of the early 1970s, which sent many western economies into a tailspin. Four decades later that issue remains one of our most pressing concerns. Who, then, will manage the necessary solutions? Who will take charge? Leadership can be a terrifying threat to the freedoms of ordinary people excluded from the chain of command, or it can be the knife which cuts through confusion and seeks out the common good. Webb's case is fascinating, for his career as NASA chief exemplified the unnerving power of large-scale governmental manipulation, even as it showed the wonders that could be achieved when so many people worked towards solving a particular challenge.

Of course, when we tell stories about the past, we have a tendency to focus on just a few key individuals, such as Webb and von Braun or their brave Russian rival Sergei Korolev. History books are replete with kings whose biographies are an easily digestible shorthand for how their entire kingdoms behaved. We read about the tactics of generals who led grand armies, because we cannot sensibly absorb the stories of every soldier and every last skirmish on the battlefield. Likewise, all tales of the space age must be subject to these limitations, for by the mid-1960s, during a brief few years at least, more than half a million people were working somewhere in or around the Apollo project, and all of them were essential.

NASA's administrators spent most of their time handling the politics and the broader daily management challenges of space, and relied on many other colleagues to deal with the finer engineering details. A valid argument can be made that Webb's immediate deputy, Robert Seamans, ran NASA at a day-to-day level; and until illness called him away, Hugh Dryden was NASA's wise elder counsel and co-administrator. Other engineer-managers in the vast Apollo team became familiar to the public in the 1960s, for a few years at least, such as Robert Gilruth, George Mueller, Sam Phillips, Joe Shea, Gene Kranz, Rocco Petrone, and Max Faget. Neil Armstrong, Buzz Aldrin, and Michael Collins were just the tip of a colossal pyramid of human endeavor and dedication. The significant legacy of Apollo is not technological but moral. NASA's staffers were unafraid to accept responsibility for their appointed tasks, whether they resulted in luminous success or horrific failure. Part of the reason for their fearlessness is that they were trusted to do their jobs, and trust tends to bring out the best in people. This lesson, above all, needs to be retained from the lunar adventure. As we send Apollo's machines to the museum, we should keep alive the memory of its people.

5 RETURNING TO THE MOON

RETURNING TO THE MOON

5

IF NASA WANTS TO REVISIT THE MOON by the year 2020, as it has recently promised, then it must hope to find staffers of similar drive and determination, then relearn how they did things in the 1960s and do likewise again. But times have changed, and it is no longer obvious that government agencies are best equipped to create the technology for space adventures. Half a century ago, keen-eyed politicians could identify useful areas for national research, such as aviation, computing, rocketry, and nuclear energy. Today it's all anyone can do to just keep in touch with the bewildering pace of developments in medicine, genetics, electronic consumer goods, personal computing, and global communications. Politicians find it difficult enough to respond to the social and environmental impact of the myriad technologies that are already on the loose, let alone urge the invention of new ones.

In the 1960s, NASA was at the vanguard of a new technology. Today it may be the custodian of an old one. Rockets are no longer one of the defining machines of our times. The ballistic missiles that heralded the space age have lost their edge in an era of "asymmetrical warfare." A few fanatics with ten cent craft knives can change the world in a day, while the expensive nuclear megatons in their underground silos are all but impotent now that our dread of nuclear weapons has been overtaken by our fear of terrorism. But none of these pessimistic thoughts can take away the glory from Apollo. It reached the moon in the fulfilment of a dream that had enticed humankind for many generations; and the scale and ambition of NASA in those years was an expression of political idealism from an America rather different to the one we know today: a nation that hoped for a better future rather than simply trying to protect itself from the present. Some people would like to revive that spirit, at least in space affairs.

NASA has some rebuilding of trust to do. On January 28 1986 space shuttle Challenger lifted off into a freezing winter sky, and seventy-three seconds into the flight it blew apart. All seven crew members were killed: Commander Dick Scobee and his co-pilot Mike Smith; mission

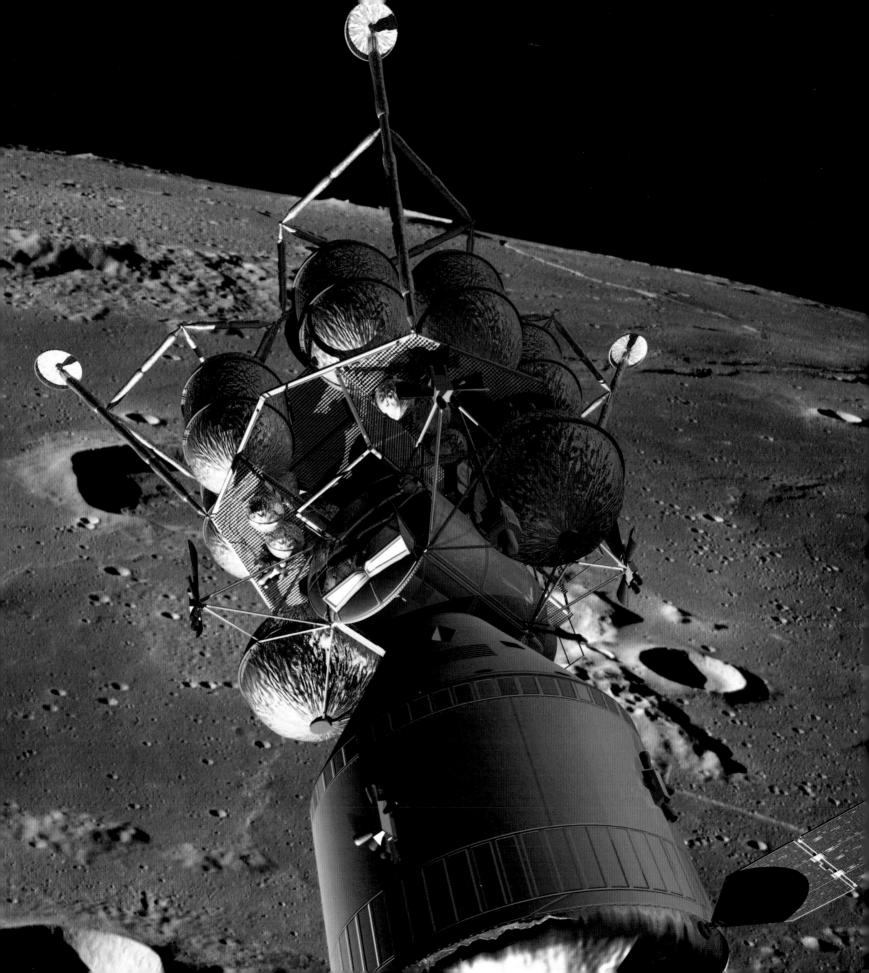

specialists Judy Resnick, Ellison Onizuka, Ron McNair, and Greg Jarvis; and a young teacher launched as part of a schools' education project, Christa McAuliffe. Today we can compare and contrast the 1967 Apollo accident with the Challenger's loss. The people of Apollo were caught off-guard by something that they had never encountered before, a dramatic fire in a spacecraft sitting quietly on the ground. The shuttle disaster was flagged by earlier failures, and in particular, repeated instances of partially burned rubber seals in boosters: exactly the problem that doomed the Challenger. Taxpayers are, perhaps, willing to forgive such disasters, but only if NASA can be said to have acted in good faith to try and prevent them.

A MAVERICK INVESTIGATOR

After the Challenger accident, the renowned physicist Richard Feynman was asked to join the board of inquiry, known as the Rogers Commission. Frustrated by the dull and pointless verbiage of due process, he staged an unauthorized little drama for the TV cameras that forced the pace. He dipped a piece of rubber into a glass of iced water and showed how it hardened when it got cold. "Do you think this might have some relevance?" he asked, knowing very well that it did. He'd created a memorably vivid demonstration of the notorious "O" ring leak that had destroyed the shuttle during a freezing winter launch.

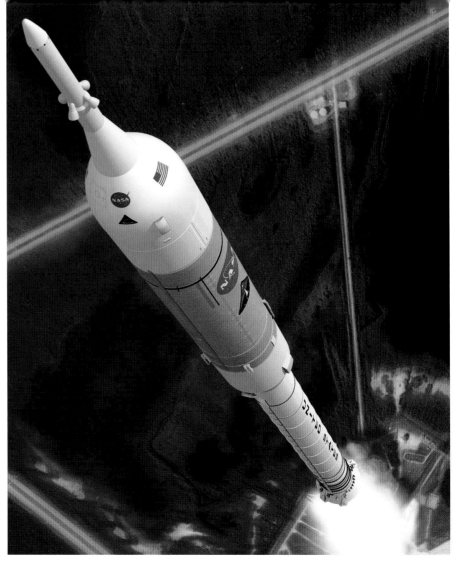

A shuttle-derived solid rocket booster sends the Orion capsule into orbit.

Feynman wanted to probe further: "If NASA was slipshod about the leaking rubber seals on the solid rockets, what would we find if we looked at the liquid-fuelled engines and all the other parts that make up a shuttle?" He was told the inquiry wasn't briefed to look at those, because no problems had been reported; so he made unauthorized trips to several NASA field centers and did some digging. As a very charming and sympathetic physicist, he had little difficulty winning the trust of staffers at ground level. "I had the impression that senior managers were allowing errors that the shuttle wasn't designed to cope with, while junior engineers were screaming for help and being ignored."

Finally, he turned to the complex relationships between the space agency's many departments and their suppliers: "NASA's propulsion office in Huntsville designs the engines, Rocketdyne in California builds them, Lockheed writes the instructions, and NASA's launch center in Florida installs them. It may be a genius system of organization, but it seems a complete muddle to me." In the last days of the inquiry, he made a plea for realism. "For a successful technology, reality must take precedence over public relations, because Nature cannot be fooled."

Other commentators identified a malaise within NASA, where a bureaucracy that was created to get a specific job done—going to the moon—had become, instead, an organization whose only purpose is to ensure its own survival. That crisis stemmed from a lack of strong, committed leadership in the setting of goals, both from within NASA and the White House. If the people in an organization have a goal to believe in, they will work cooperatively. If that goal is lacking, then people start worrying about simply keeping hold of their jobs. They will then compete with each other, internally, and the organization will fail.

On February 1 2003, a second space shuttle, Columbia, was destroyed when it disintegrated during re-entry. The entire crew of seven astronauts died. An investigation over the ensuing months concluded that a suitcase-sized chunk of foam insulation from the huge external fuel tank had peeled off during launch. It slammed into the front of the shuttle's left wing, making a small but ultimately lethal hole in the heat-resistant panels. This disaster, horribly echoing the Challenger disaster of 1986, forced America to reconsider its entire space effort.

The Columbia Accident Investigation Board (CAIB) concluded that it is inappropriate to carry astronauts in the same part of a spacecraft that contains the rocket engines, because of the risk of explosions, collisions, or aerodynamic break-up of the structure. The Board noted that Apollo's tough and compact crew compartment could always be instantly separated from other modules or booster rockets. When a section of the Apollo 13 spacecraft exploded on the way to the moon in April 1970, the rear service module with the rocket engine was torn open, but the crew capsule itself was unharmed, and it returned safely to the earth. The shuttle has no such self-contained module capable of separating its crew from the rest of the machinery

A full-scale dummy version of the new Orion capsule arrives at a NASA center.

in the event of a problem during flight. The CAIB also recommended that future crew modules for any spacecraft should be carried on the uppermost tips of their carrier rockets, so that no launch debris can fall onto them from any hardware above, as had happened in the Columbia disaster (and, to a less destructive yet still alarming degree, during many other shuttle missions too). The CAIB also urged that NASA's remaining shuttles, Atlantis, Endeavor, and Discovery, should be retired from service by 2010. But what should replace them?

A heavy-lift launch vehicle, aided by solid rocket boosters inherited from the space shuttle system, hauls the new lunar landing craft into earth orbit.

NEW PLANS FOR THE MOON

After deliberating with his advisors for many months, President George W. Bush made a tele-vised speech in January 2004 during a special visit to NASA headquarters in Washington. "Today I announce a new plan to explore space and extend a human presence across our Solar System." The first goal, he said, was to return the shuttles to flight status, so that they could participate in the long-delayed completion of the International Space Station.

No surprises there. But as his speech continued, radical new ideas emerged: "Our second goal is to develop and test a new spacecraft, the Crew Exploration Vehicle, by 2008, and to conduct the first manned mission no later than 2014. Our third goal is to return to the moon. Using the Crew Exploration Vehicle, we will undertake extended human missions to the moon

Launched separately on a smaller rocket, the Orion crew capsule docks with the lunar lander, attached to its heavy-lift rocket. The upper stage of that larger rocket then fires again to carry the combined spacecraft towards the moon.

as early as 2015, with the goal of living and working there. With the experience and knowledge gained on the moon, we will then be ready to take the next steps of space exploration: new human missions to Mars and to worlds beyond." He summed up by saying:

In the past thirty years, no human being has set foot on another world, or ventured farther into space than 386 miles, roughly the distance from Washington, DC to Boston, Massachusetts. America has not developed a new vehicle to advance human exploration in space in nearly a quarter-century. It is time for America to take the next steps.

The Crew Exploration Vehicle (CEV) otherwise known as Orion, is a reversal of much that NASA has worked towards over the past three decades. The dream of an all-purpose shuttle, flying cheaply and regularly like a commercial cargo plane, never came to fruition. Every time that a shuttle puts a satellite or a space station module into orbit, the costs are exaggerated because astronauts always come along for the ride. The launch weight given up to life support and crew cabins limits the amount of useful payload a shuttle can carry in its cargo bay. It would be more efficient, say Orion's designers, to haul cargo in uncrewed rockets, and then launch the astronauts separately in a small module that carries people and nothing else. There's no need for the shuttle's huge wings and cargo bay doors. All that's left to worry about is the crew cabin at the front. Replace that with a capsule that fits on the end of a conventional rocket, and it can serve as an escape pod if anything goes wrong. It all looks very familiar. Robert Seamans, NASA's Deputy Administrator in the Apollo era, was one of a dozen ageing veterans called upon to advise Orion's designers:

I served on what they called the "Graybeard Committee," all these old hands who knew how we'd reached the moon the first time around. Astronaut John Young was there, and he'd flown in Gemini, plus two Apollo missions, and he'd commanded the first shuttle flight. We didn't fool around. What we came up with is very similar to what we did with the Apollo capsule.

After more than three decades, it is impossible to recreate Apollo's giant Saturn rockets and launch gantries. All the factory tooling was scrapped, and even the original drawings have been lost. Instead, the Orion capsule will be launched on top of a solid rocket booster, as currently employed by the shuttle. A lunar landing craft, the Altair, will be carried into orbit, unmanned, aboard a heavy-lift rocket. The Orion and its crew will locate it in orbit, make a docking, and head to the moon. On completion of the outward journey, Altair will detach and drop down to the surface. At the end of its mission, the lower part of the craft will stay behind, while the crew module blasts back into lunar orbit and makes a rendezvous with Orion, which then heads back to Earth for an Apollo-style parachute landing.

> "Don't tell me that man doesn't belong out there. Man belongs wherever he wants to go—and he'll do plenty when he gets there."
>
> **Wernher von Braun, 1958**

Is this progress, or a leap back in time? Aerospace engineer and writer Robert Zubrin has worked tirelessly over the past decade to promote a human Mars mission. (His lobbying group, the Mars Society, has some 8000 members.) The wayward logic of space planners sometimes exasperates him. "In the 1970s, when President Nixon killed the Apollo program and ended lunar exploration, NASA said we would do all that again one day, after we had developed cheaper transportation to orbit using a winged shuttle. We've spent thirty years trying that, but the shuttle has cost us more to fly than the Saturn boosters that we had for the lunar flights. Now the shuttle is going to be replaced with a little capsule just like the Apollo."

A NEW SPACE RACE?

Perhaps that's not such a terrible thing. A modern replay of Apollo might seem like a brand new adventure to the very many people on this earth who were not even born when Armstrong and Aldrin touched down on the Sea of Tranquility. But can we be sure that America will be in charge this time around? There is a politically sensitive gap between the year 2010, when the shuttle is retired, and 2015, when the Orion makes its debut. Other countries will not miss an opportunity to try and fill the gap, both technically and in terms of global prestige.

The "back to the future" technology for Orion is shaping up fast, but there is no guarantee that it won't be cancelled before it reaches the moon. NASA marks its many astonishing achievements over the last half-century with budgetary doubts clouding its future. John Logsdon, the influential head of the Space Policy Institute in Washington DC, has warned that everything could change at a moment's notice, subject to the whims of Congress. "If the proposals for

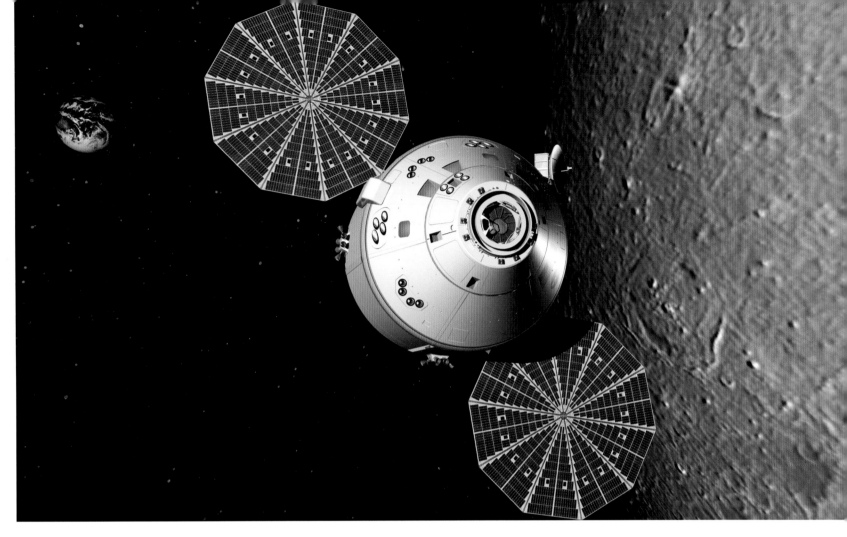

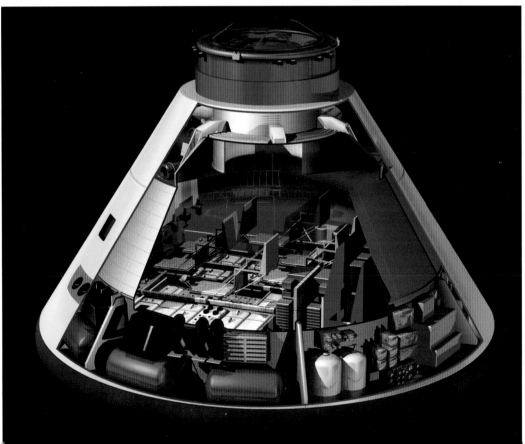

The Orion capsule for a return to the moon (above and right) has been described by some NASA insiders as "Apollo on Steroids."

Engineers test an experimental lunar rover. Each set of wheels can pivot in any direction, giving the vehicle the ability to drive sideways, as well as forward or backward.

human space exploration fail politically, then major initiatives that depend on government spending could lose momentum. It's far from certain that we're going back to the moon, let alone on to Mars. I think we are still struggling to answer the most basic question: why is the government in the business of sending people into space?" Maybe Logsdon's pessimism is well-founded, but history tells us that if the White House senses anyone else bidding for the moon, NASA may be called on to deliver another stunning victory.

China has begun its own human spaceflight program, launching missions at the rate of one every two years: a cautious rate which, the Beijing authorities say, will accelerate over the next decade. In all likelihood, political tensions will prevent China from playing any role aboard the International Space Station. The same cannot be said of Russia, Japan, or the European Space Agency ESA. Together they are developing the Crew Space Transportation System (CSTS). It should be capable of reaching lunar space by the end of the next decade. CSTS will exploit technology from a ship that ESA has already built and flown. The Automated Transfer Vehicle (ATV) delivers eight tons of supplies and equipment to the Space Station. Launched to orbit by ESA's Ariane V rocket, but carrying its own independent propulsion and guidance systems, the ATV has a pressurized cargo module designed for crew access over a period of several months during its attachment to the station. It therefore already satisfies most of the criteria for human spaceflight. The Russian contribution to the CSTS will be their many years' experience of bringing crews back to earth by the use of re-entry capsules. Meanwhile, private companies, such as Constellation Services and Space Adventures, want to revive Russia's 1960s project to send cosmonauts around the moon without landing on it. Upgraded Soyuz machinery could accomplish this, so long as investors are prepared to put up $100 million. Such companies are, perhaps, about to transform a wide range of space activites, both in earth orbit and on or around the moon—and one day, perhaps far beyond.

Certainly these companies—and the potential international alliances between different space agencies—have a great deal to contribute in terms of hardware and new devices, just so long as NASA and its political overseers in Congress allow this to happen. The next generation of crewed lunar landing craft will be larger and tougher than the Apollo modules from the 1960s. Astronauts will live aboard them for several weeks at a time, and inflatable modules could be attached to the airlocks after touchdown, adding even more room. Permanent living quarters may eventually be constructed under protective blankets of lunar topsoil. Pressurized wheeled rovers, the ultimate "off-road" vehicles, will allow astronauts to travel great distances in search of scientific data. The rovers, living modules, and other lunar hardware will be designed so that the technology for a Mars mission can be derived from them in the future. In this new scenario, no ideological clash exists between human and computer-controlled exploration. An extended family of robotic tools will support astronauts rather than competing with them: long-range scout rovers, mapping orbiters, communications relays, cargo ferries, energy supply systems, and astronomical instruments.

Since future astronauts will stay on the moon for many weeks or even months at a time, rather than just a few days, the question arises: where on the moon should they live? In full daylight the lunar surface can reach blistering temperatures of 100°C, while at night it plunges to -180°C. This environment will place great stresses on a habitable base. The long nights (two terrestrial weeks) would also play havoc with solar power supplies. Luckily, there is at least one location that remains permanently in daylight, while benefiting from a relatively benign temperature all year round of -50°C: not much worse than the temperatures we encounter in the Antarctic, and perfect for a base designed for long-term habitation. One magic spot is on the

The crew capsule will touch down on land, with the help of airbags, or at sea.

rim of Peary crater, close to the lunar north pole. Other useful sites may exploit the opposite characteristic: permanent shadow. Since the dawn of space flight, many astronomers have wanted to build optical and radio telescopes inside deep lunar craters where they are shielded from solar light pollution or unwanted "noise" from earth's countless radio sources.

Whatever happens, if we do choose to go back to the moon in the next decade, it should not just be to conquer a new world and plant shopping malls in the lunar dust. We should return there to discover more about ourselves, and to look back, with wiser and more appreciative eyes, at the world we leave behind. In fact, that was one of Apollo's greatest achievements. When Apollo 8 flew around the moon during Christmas 1968, the three astronauts aboard took color photographs of the earth rising above the lifeless lunar horizon. Our home world seemed so vulnerable, drifting alone in the forbidding blackness of space.

Those pictures helped inspire the environmental movement—and it was important that human eyes had seen that sight, not just the cameras. Apollo 8 astronaut Bill Anders said, "I trained hard for many years to fly around the moon and take a close look at it, because I thought that's what the mission was all about. When we finally got home, we realized that we had discovered something much more precious out there: the earth. That alone was worth the relatively few billions of dollars that we spent on Apollo."

Perhaps it's always the earth that we are looking for. In his 1961 novel *Solaris*, the Polish science fiction author Stanislaw Lem wrote, "There's nowhere we cannot go. And in that belief, we set out for other worlds, all brimming with confidence. And what are we going to do with them? Rule them, or be ruled by them. This is the only idea in our pathetic minds. What a useless waste!" Lem understood that space is just a larger than usual screen onto which we project our earthly fears and desires. We always look for something recognizably human in those alien places. And when we do encounter the genuinely alien—the deep cold, the dark and airless infinities, the terrifying loneliness of the void between stars—we cannot understand these things, and we return from our celestial voyages slightly disappointed.

> "I don't think that the human race will survive the next thousand years, unless we spread out into space. Eventually, we will reach out to the stars."
>
> **Professor Stephen Hawking, 2008**

The stars are not some kind of a heaven that we might reach if only we build the appropriate machines. The stars are just stars, and space is just another physical volume for us to travel through and occupy; and as the cultural critic Stanley Kaufman tried to remind audiences as they watched Apollo 11's grainy television pictures, our essential human experience of space is limited to the view through a window, a camera, or a helmet visor: "Space only *seems* large. For human beings, it is confining. That is why, despite the size of the starry firmament, the idea of space travel gives me claustrophobia." Part of us rebels against this thought, but Kaufman was probably right. Our idea of space is an illusion—a wish more than a truth.

Stanley Kubrick's 1968 movie *2001: A Space Odyssey* was certainly in love with the cosmic hopes and expectations generated by Apollo. His invisible, apparently benevolent aliens guided our fate. But the movie was just a piece of fiction. In the end, as Kubrick recognized, we have to find our own solutions. We can't rely on space travel:

The most terrifying fact about the universe is not that it is hostile, but that it is indifferent. But if we can come to terms with this indifference and accept the challenges of life within the bounds of death, our existence as a species can have genuine meaning. However vast the darkness, we must supply our own light.

Discoveries in space are only half the story. It's what we make of them that counts. As always, human progress can only come from inside us. We can't find it "out there," no matter how far we travel. Forty years after Lem's *Solaris*, the British author J. G. Ballard expressed a similar sentiment. "The biggest developments of the immediate future will take place not on the moon or Mars, but on earth, and it is inner space, not outer, that needs to be explored. Even in space, the most alien creatures we'll confront are ourselves."

We will find in the depths of the cosmos whatever we bring along in the first place. Let's make sure we find something wonderful.

"Looking back, we were really very privileged to live in that thin slice of history where we changed how we look at ourselves, and what we might become, and where we might go. So I'm very thankful that we got to be a part of that."

Neil Armstrong, 2001

6 APOLLO 11 MISSION IMAGES

The crew of Apollo 11 are assisted into their spacesuits, then leave for the launch pad.

The crew enter the base of the pad, then take a lift up to the capsule.

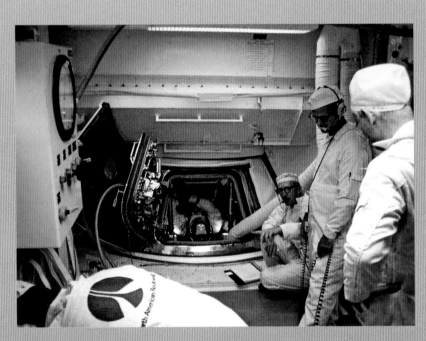

The crew are inserted
into the command module (left).

Launch controllers (below) watch
the Saturn V's ascent (right) from
the firing room at the Kennedy
Space Center, while Lyndon Johnson
(overleaf) watches from the VIP
observation stands.

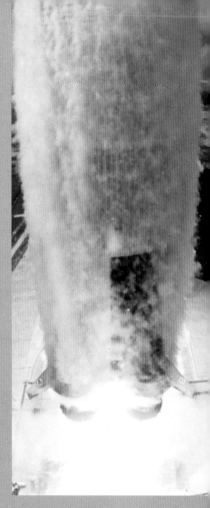

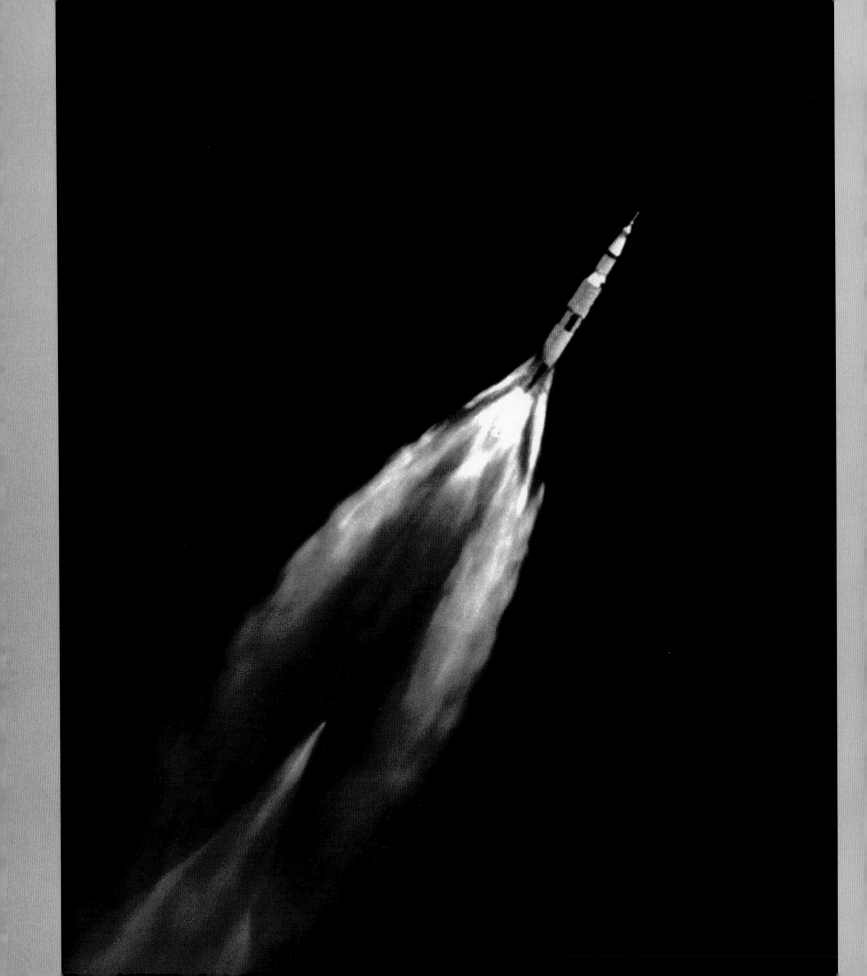

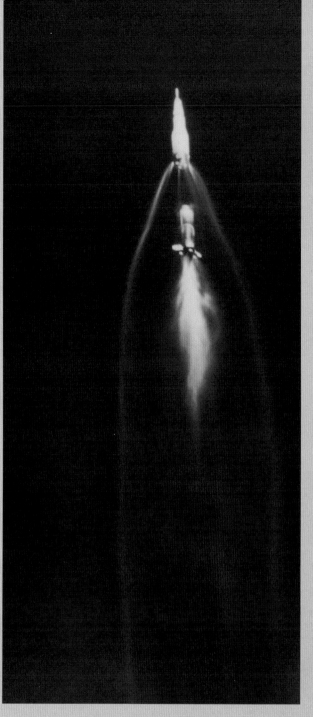

The Saturn V ascends on a plume of flame, then drops its spent first stage. From the moment of lift-off, the entire ascent to earth orbit takes just nine minutes and thirty seconds.

"We shall send to the moon, 240,000 miles away from the control station in Houston, a giant rocket more than 300 feet tall, made of new metal alloys, some of which have not yet been invented, capable of standing heat and stresses several times more than have ever been experienced, fitted together with a precision better than the finest watch, carrying all the equipment needed for propulsion, guidance, control, communications, food, and survival, on an untried mission, to an unknown celestial body, and then return it safely to earth, re-entering the atmosphere at speeds of over 25,000 miles per hour, causing heat about half that of the temperature of the sun."

John F. Kennedy, 1962

Apollo 11 enters earth orbit.

Neil Armstrong during the flight.

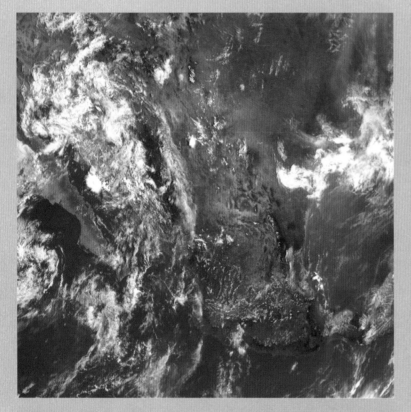

"This was a project in which everybody involved was, one, dedicated, and, two, fascinated by the job they were doing."

Neil Armstrong, 2001

As earth recedes, the command module swivels around to connect with the lunar module.

The lunar module's fragile outer skin is evident in this view.

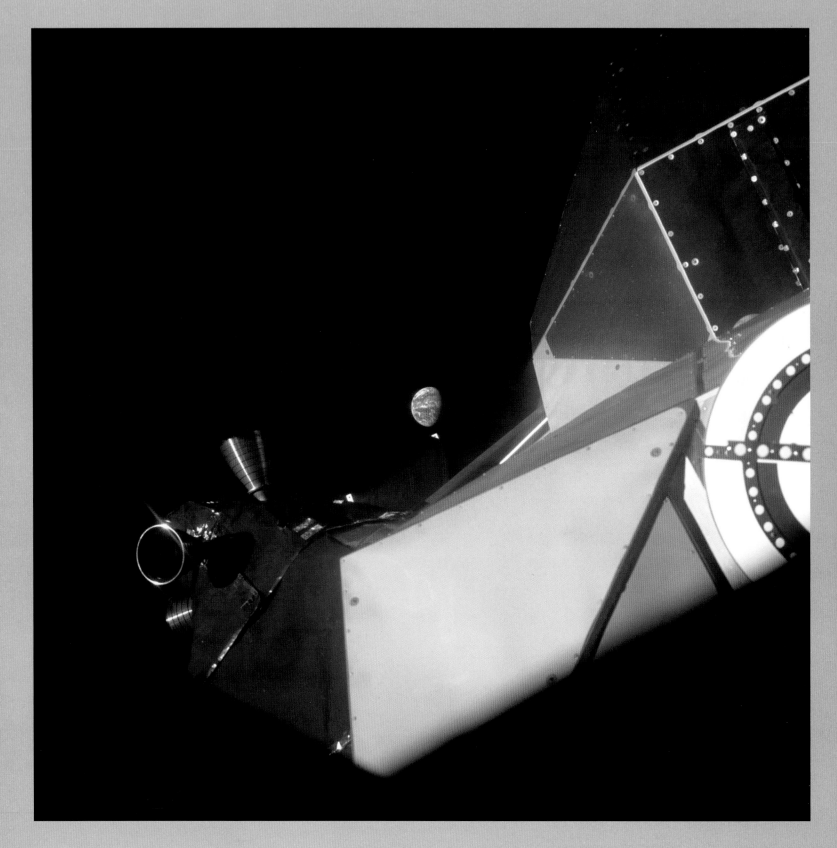

The earth dwindles to a disc so small that the astronauts can hide it behind a thumb.

Testing out a television camera inside the docking tunnel linking the two ships.

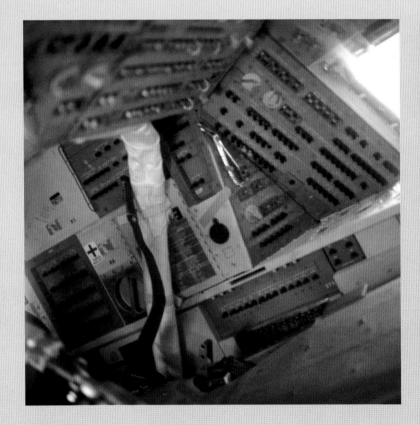

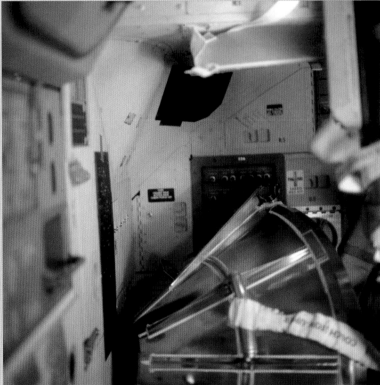

"The assistance that we receive from our technological devices is countered by the restrictions that they place upon us. The very same scientific knowledge that constructs our spacecraft also informs us of apparently insurmountable physical limits."

Charles Lindbergh, 1974

Cramped mechanisms: the disassembled docking probe, and (top left) the maze of switches in the capsule.

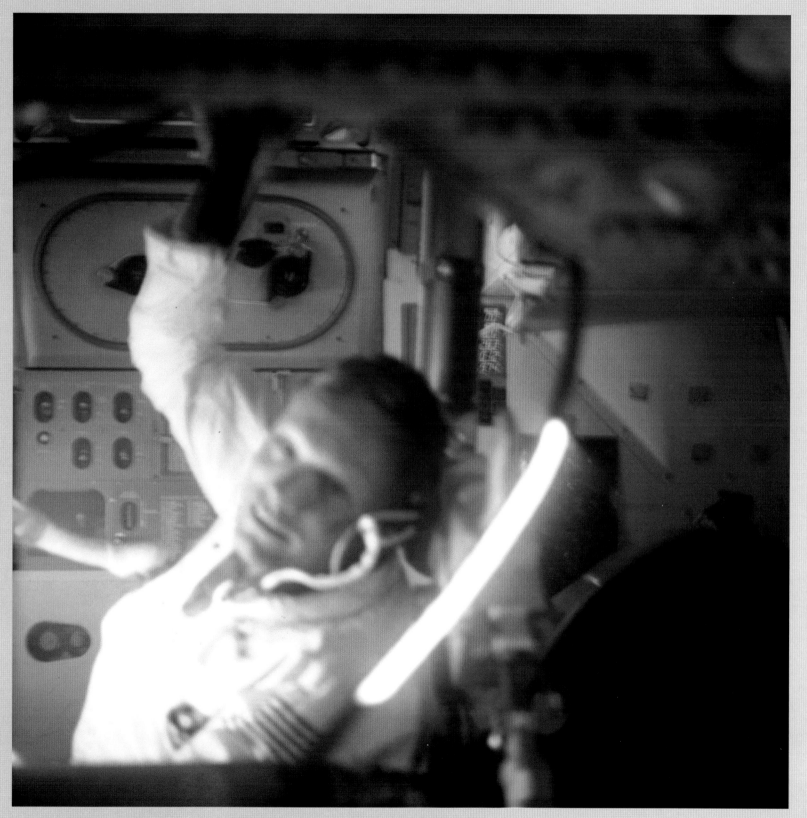

Another shot of Armstrong mid-flight. The white streak is random glare from the interior lighting.

A confident-looking Aldrin, photographed by Armstrong.

"I had been to the moon. What could I possibly do next? I suffered from what the poets have described as 'The melancholy of all things done.' Without some new goal, I was aimless."

Buzz Aldrin, 1973

Aldrin's work station on the right-hand side of the lunar module's cabin.

The moon approaches.

The moon can be a hostile place to visit—yet we persist in wanting to go there.

"I can remember at the time thinking, 'There's got to be a better way of saying this.' If you're going to tell three human beings that they can leave the gravitational field of earth, what are you going to say? You'll invoke Christopher Columbus, or a primordial reptile coming up out of the swamps onto dry land for the first time, or you'll go back through the sweep of history and say something meaningful. Instead, all we had was our technical jargon."

Michael Collins, 1997

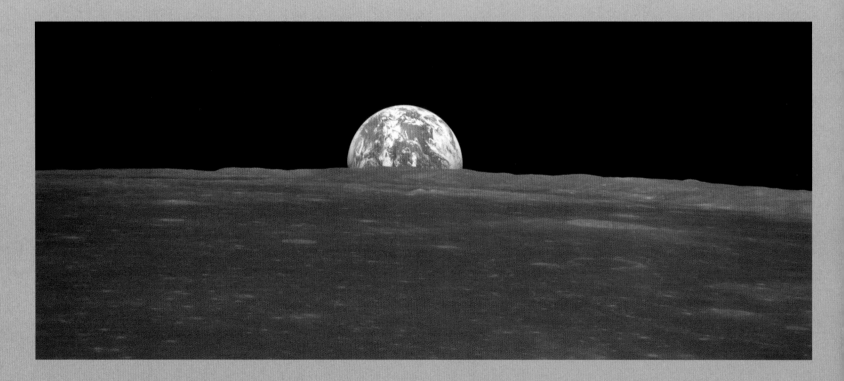

"The earth looks so small. You see an ocean and gaseous layer, just a tiny bit of atmosphere around it, and compared with all the other celestial objects, which in many cases are much more massive, more terrifying, it just looks like it couldn't put up a very good defense against a celestial onslaught."

Neil Armstrong 2001

As Apollo 11 completes its first orbit of the moon, the blue earth rises over the horizon.

"I think a future flight should include a poet,
a priest, and a philosopher.
We might get a much better idea
of what we saw."

Michael Collins, 1974

The lunar module undocks in preparation for the landing descent.

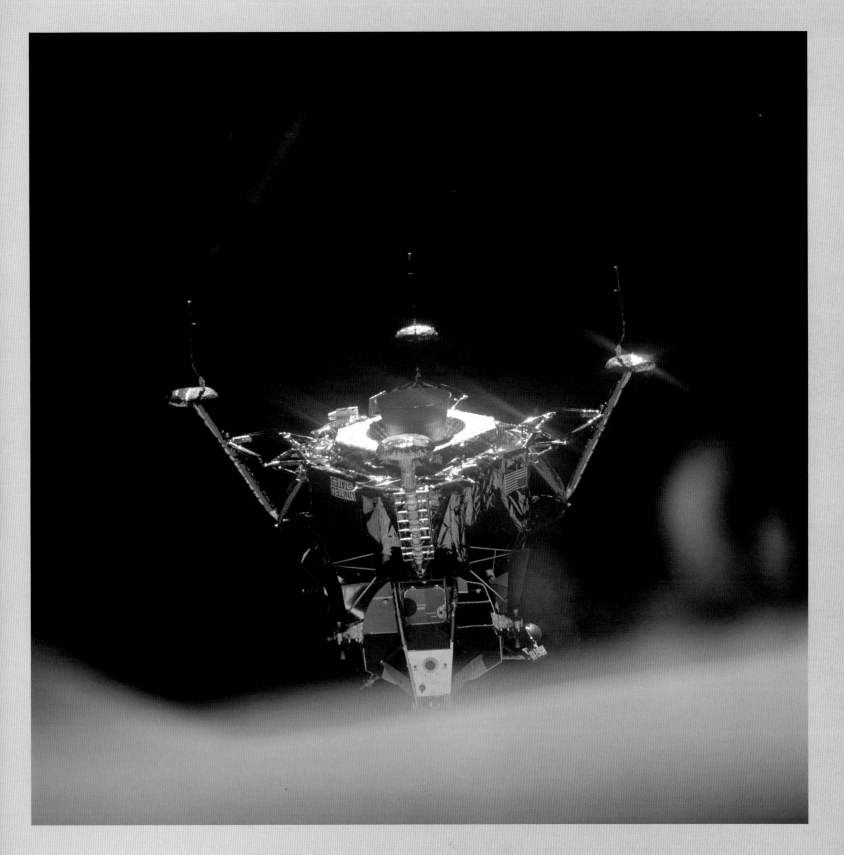

The notion of "upside-down" doesn't mean much in space.

Everything is in the detail: small rocks, craters, and other hazards await.

Touchdown on the Sea of Tranquillity.

Bootprints and the flag.

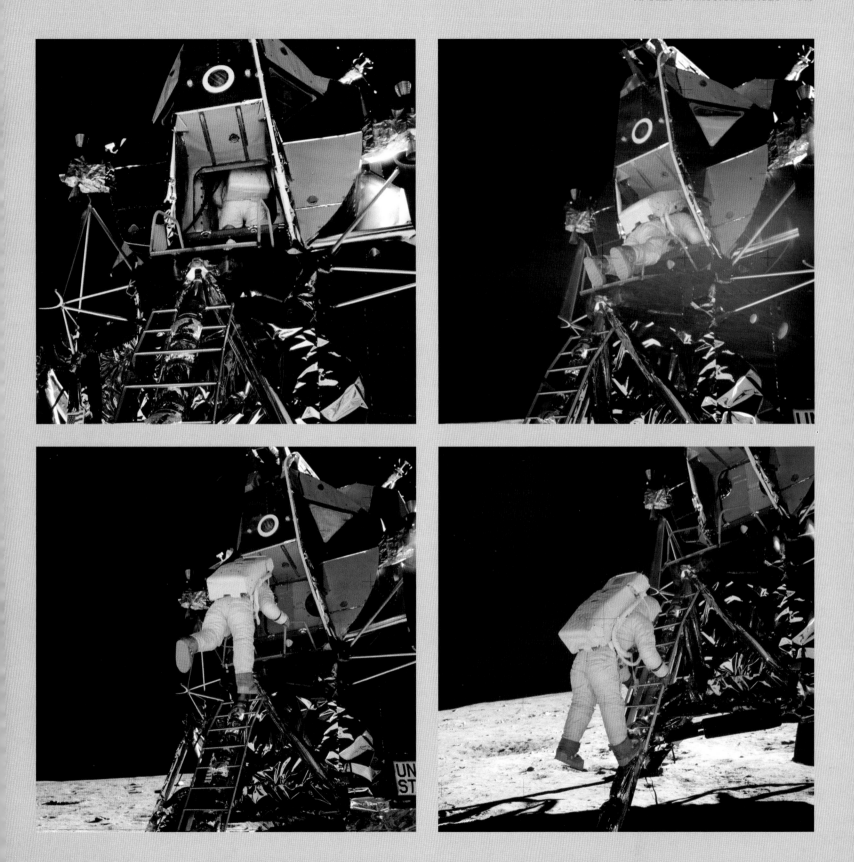

Aldrin is the second man on the moon, but the first to be photographed there.

Aldrin is safely down—without tripping: a fear for all the lunar astronauts.

The commemorative plaque on the lunar module's front landing leg.

"I just don't deserve the attention for being the first man on the moon. Luck and circumstance put me in that particular role. That wasn't planned by anyone."

Neil Armstrong, 2005

Ghostly TV pictures of the moon walk, relayed around the world via NASA's mission control.

Aldrin salutes the American flag.

"I expected the unexpected, and I went with an open mind. I think that the visual scene was best described by my words on first landing: 'magnificent desolation.' It was magnificent because of the achievement of being there, and desolate for the eons of lifelessness."

Buzz Aldrin, 1998

These two photographs are the only ones to show Neil Armstrong on the lunar surface.

The most famous picture in space history: Buzz Aldrin on the moon.

Aldrin deploys a foil sheet to collect subatomic particles streaming from the sun.

These bootprints are still on the lunar surface today, undisturbed by wind or rain.

"The tools of the Apollo project were not developed for people to live on the moon on a long-term basis. Therefore, economically and politically, they could not be sustained and supported once all the original missions had been completed."

Buzz Aldrin, 1998

Shots of the lunar module, showing its hand-made construction.

One of the footpads sitting in the very shallow soil.

Portrait of an extraordinary machine, unlike anything that we normally experience.

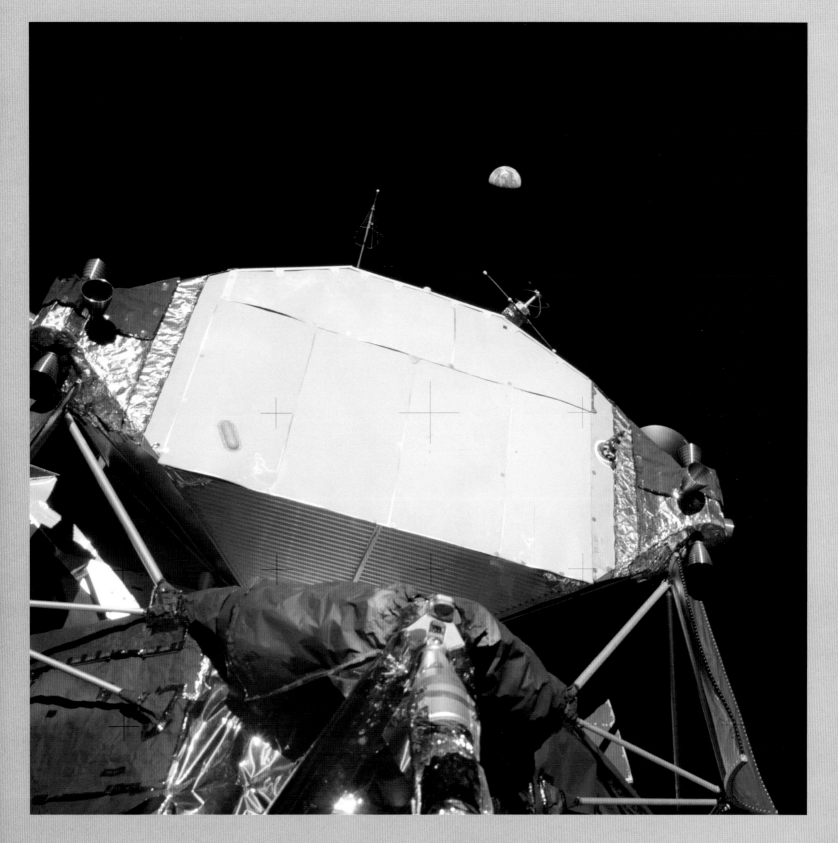

The earth seen above the rear of the lunar module.

Collecting equipment from the flanks of the lander.

Aldrin walks into the distance to deploy scientific instruments.

Aldrin deploys a seismometer to measure moon quakes.

"In those days, I did get tired of being asked the same things over and over and over and over and over and over and over again. 'What was it like?' You can see it in their eyes, 'What kind of a person is this? He's just come back from the moon, and he's not even excited about it.' It isn't the fact that I'm not excited. It's the fact that I've been asked that same question ten million times."

Michael Collins, 1979

"Magnificent desolation."

A happy Neil Armstrong, photographed by Aldrin after the successful moon walk.

"I was aware that this was a culmination of the work
of 300,000 or 400,000 people over a decade, and that
the nation's hopes and outward appearance to the rest
of the world largely rested on how the results came out.
With those pressures, it seemed the most important thing
to do was focus on our job as best we were able to,
and try to allow nothing to distract us from
doing the very best job that we could."

Neil Armstrong, 2001

Against the immensity of the moon, Apollo 11's command module can just be seen.

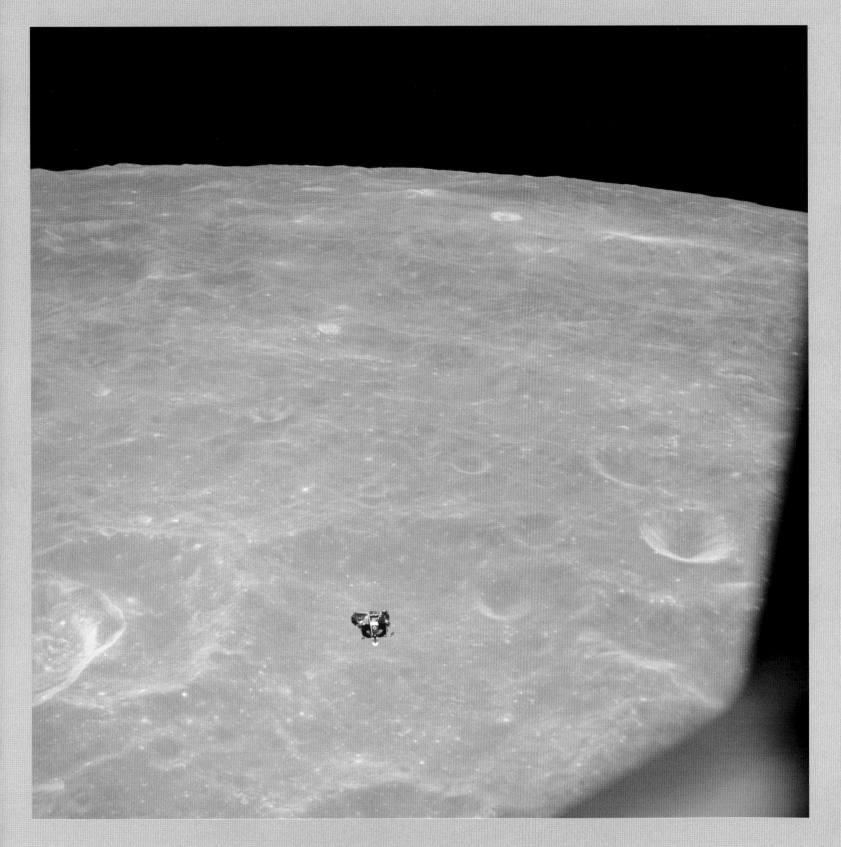

The ascent stage of the lunar module climbs back to rejoin the mothership.

Michael Collins steers the command module towards the second and final docking.

The ascent stage of the lunar module closes in for final approach.

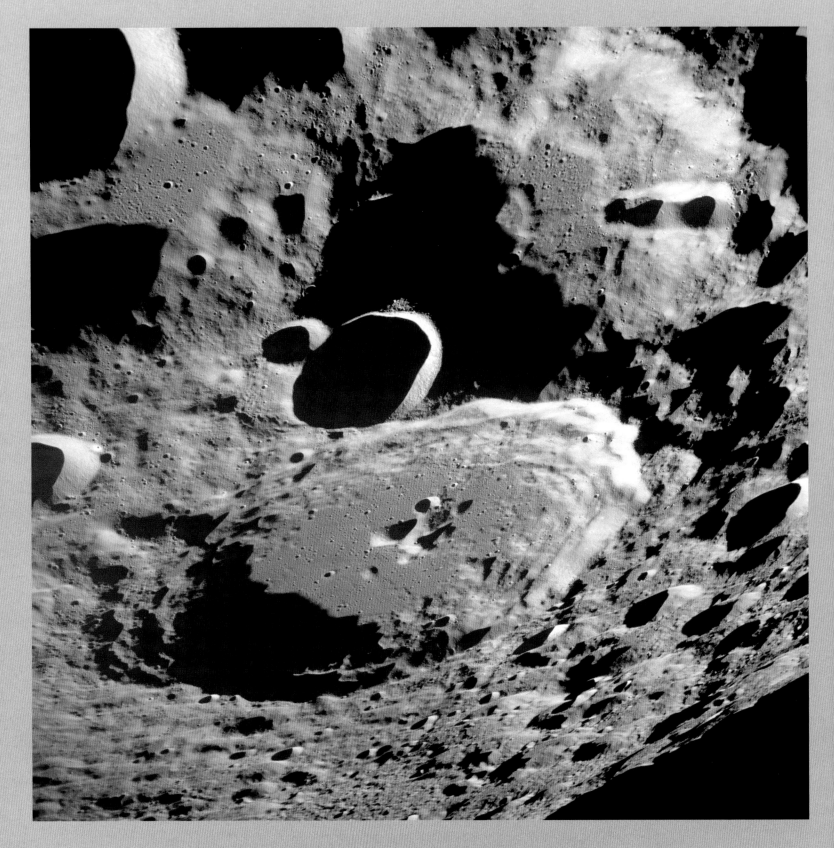

Daedalus Crater seen from orbit.

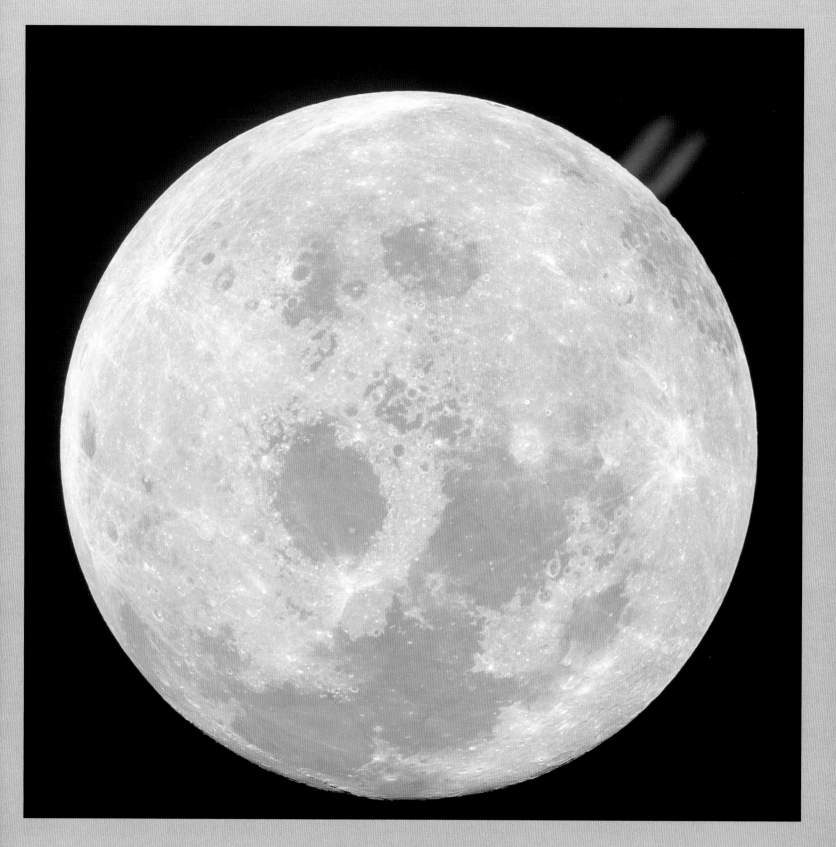

The moon recedes as Apollo heads home.

The earth looms large once again.

"Space flight is a spiritual quest in the broadest sense—one promising a revitalization of humanity and a rebirth of hope."

Buzz Aldrin, 1999

The crew of Apollo 11 in the quarantine facility.

"We set sail on this new sea because there is new knowledge
to be gained, and new rights to be won, and they must be won
and used for the progress of all people. For space science,
like nuclear science and all technology, has no conscience of its own.
Whether it will become a force for good or ill depends on man.

"Space can be explored and mastered without feeding
the fires of war, without repeating the mistakes that man
has made in extending his writ around this globe of ours.

"We choose to go to the moon in this decade, and do the other things,
not because they are easy, but because they are hard—because
that goal will serve to organize and measure the best of our energies
and skills—because that challenge is one that we are willing to accept,
one we are unwilling to postpone, and one which we intend to win.

"Space is there, and we're going to climb it, and the moon and
the planets are there, and new hopes for knowledge and peace
are there. And, therefore, as we set sail, we ask God's blessing
on the most hazardous and dangerous and greatest adventure
on which man has ever embarked."

John F. Kennedy, 1962

APOLLO 12: Pete Conrad, Alan Bean, Dick Gordon.

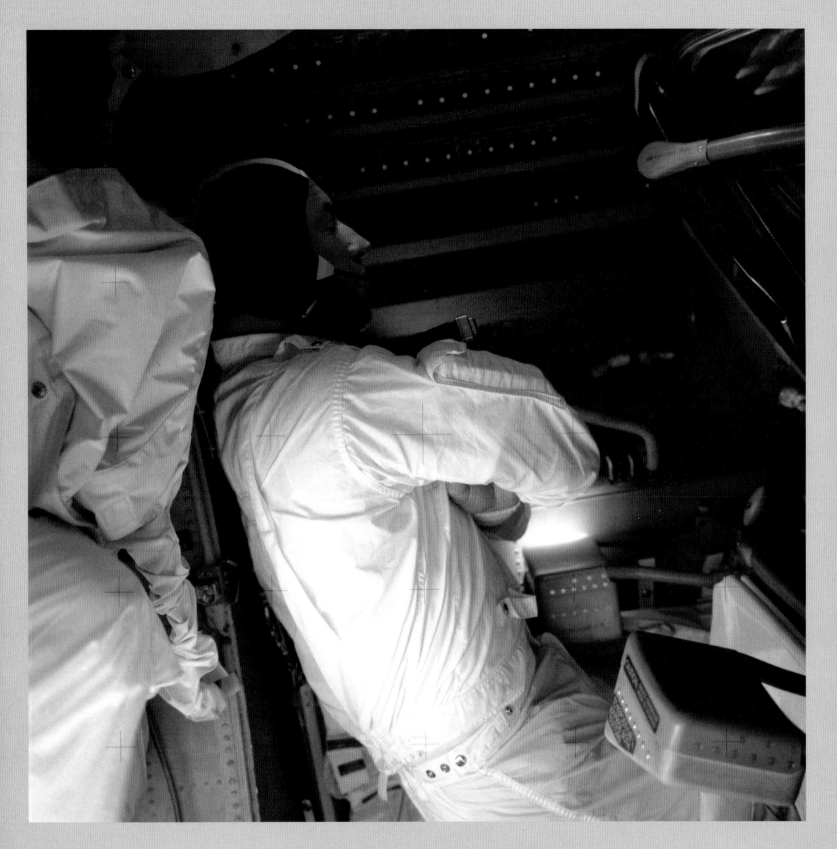

APOLLO 13: Jim Lovell, Fred Haise, Jack Swigert.

APOLLO 14: Alan Shepard, Ed Mitchell, Stuart Roosa.

APOLLO 15: Dave Scott, Jim Irwin, Al Worden.

APOLLO 16: John Young, Charlie Duke, Ken Mattingly.

APOLLO 17: Gene Cernan, Harrison Schmitt, Ron Evans.

Gene Cernan, commander of Apollo 17: the last man on the moon . . . so far.

**NASA launch controllers and other staffers turn out in force to say goodbye to the Apollo
era, as the last mission heads off to the moon in 1972 . . . and overleaf, a reminder
of what we *won't* find out in space—the richness of life on earth.**

"I trained hard for many years to fly around the moon and take a close look at it, because I thought that's what the mission was. When we got home, we realized we had discovered something much more precious out there: the earth. I believe the environmental movement was tremendously inspired by those missions. That alone is worth the relatively few billions of dollars that we spent on Apollo."

Apollo 8 astronaut Bill Anders, 1998

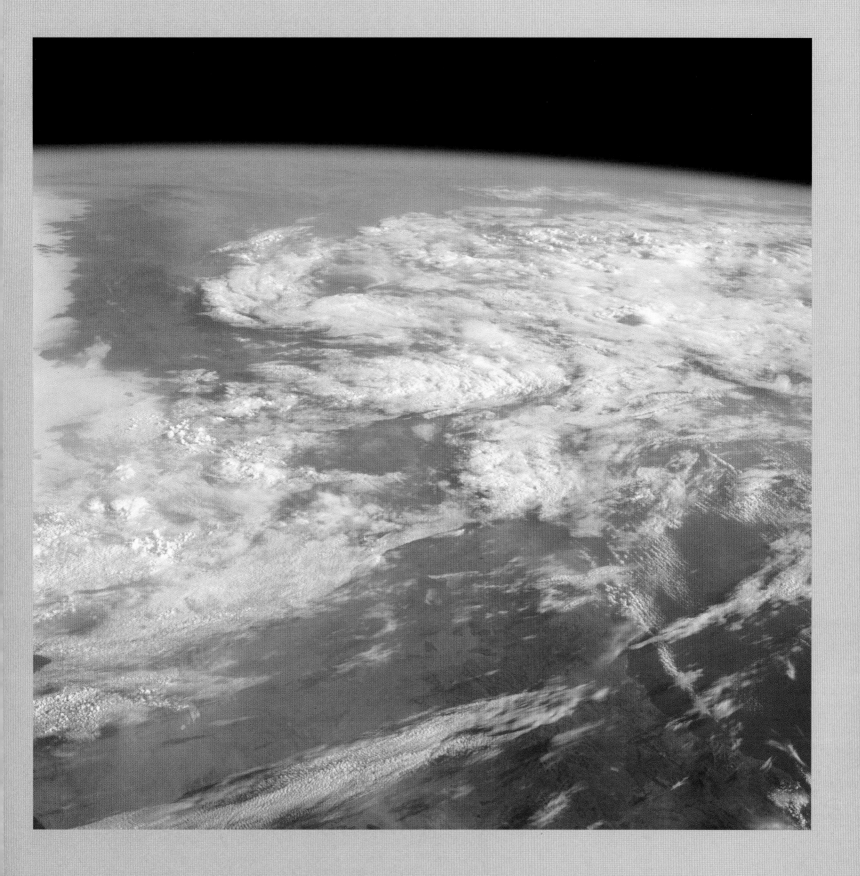

BIBLIOGRAPHY

Cadbury, Deborah
Space Race
Fourth Estate, 2005

Chaikin, Andrew
A Man on the Moon
Michael Joseph, 1994

Cooke, Hereward Lester, & Dean, James D.
Eyewitness to Space
Harry N. Abrams Inc., 1970

Harford, James
Korolev
John Wiley & Sons, 1997

Lambright, Henry W.
Powering Apollo: James E. Webb of NASA
Johns Hopkins University Press, 1995

Launius, Roger
*NASA: A History of the US
Civil Space Program*
Krieger, 1994

Mailer, Norman
A Fire on the Moon
Weidenfeld & Nichsolson, 1970

McDougall, Walter A.
*The Heavens and the Earth: A Political
History of the Space Age*
Basic Books, 1985

Murray, Charles and Bly Cox, Catherine
Apollo: The Race to the Moon
Secker & Warburg, 1989

Wolfe, Tom
The Right Stuff
Farrar, Straus and Giroux, 1979

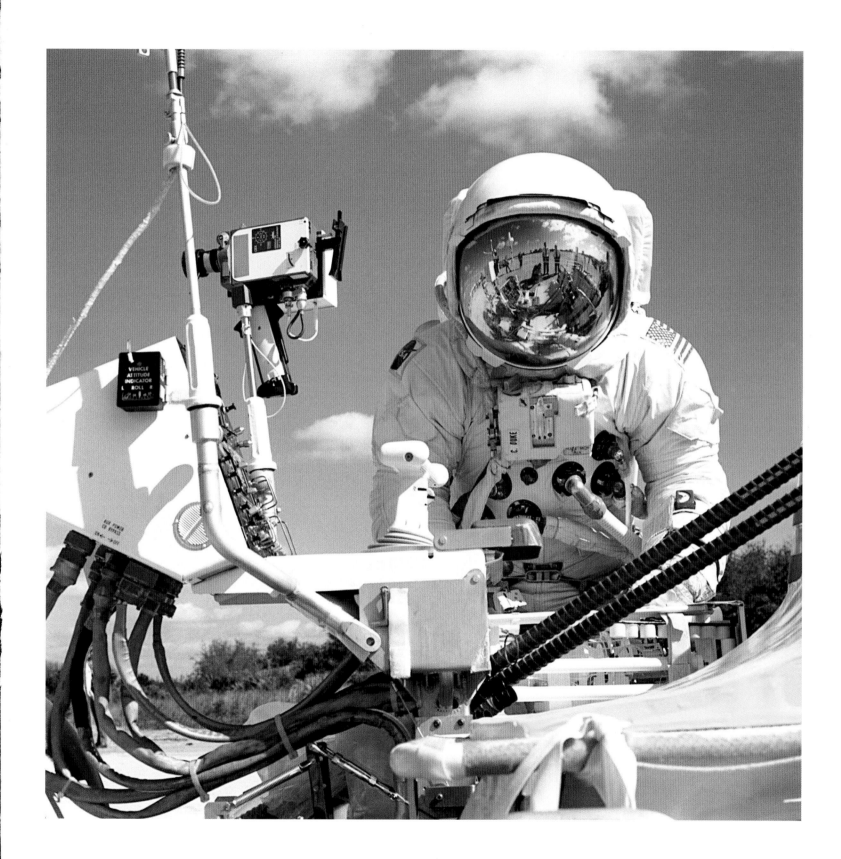

All pictures are courtesy of NASA, with the exception of:

Page 41 & 42 (top): Smithsonian Institution/Eric Long

Page 44: John Knoll

Pages 48–50, and chapter headers: John Ortmann

Page 53: Warner Brothers/Ronald Grant Archive

Page 158: Alma Haser

Additional image research and processing:

Paul Fjeld, Mike Gentry, Ed Hengeveld, Eric Jones,

Mike Marcucci, Jack Pickering, and Kipp Teague